SMOKESCREENS

THE HOODWINKING OF AMERICA

Book 1 of the Smokescreens Social Conscience Series

BY HENRY PATIÑO

Isaiah 44:6, Psalm 24:1
(Gebo Wunjo Othala Chi Rho Owns Earth)

Areli Media

Table of Contents

ACKNOWLEDGMENTS

In the early 1980s, I was privileged to meet Dr. Francis Schaeffer and his son Franky at a conference in Ft. Lauderdale, Florida. It was one of those pivotal moments in life that becomes a milestone. His words have echoed in my mind countless times as I remember the challenge he gave me. Racked with cancer, in his waning moments, so weak that he could not stand for long, this valiant man of God tirelessly continued the battle for the minds of our generation.

A group of us involved in the planning of the first evangelical national conference for the pro-life movement met in a small restaurant in Ft. Lauderdale to discuss strategy in bringing this message to the evangelical church, which at that time considered the issue a political matter that was too controversial for the church. It was, to say the least, a hot potato that few pastors were willing to juggle for fear of offending their parishioners. Included in this meeting were men such as John Whitehead, Franky Schaeffer, and Cal Thomas, whose contributions to the movement were paramount. In a private meeting afterward, Franky took some of us to meet Dr. Schaeffer in an office behind the stage of Dr. James Kennedy's Coral Ridge Presbyterian Church, which had sponsored the conference.

At the end of the meeting, Dr. Schaeffer gave us this charge: "I am close to my end now; I have done what God has given me the grace to accomplish. It is up to you now, the next generation, to stand on my shoulders and continue in the cause."

I have stumbled more than once along the way. I am not worthy to stand on the shoulders of my mentor, and I certainly do not presume to be in any way his successor. But by the grace of our forgiving Father, I am attempting to follow his charge to reach the minds of those entering into our third millennium after the incarnation.

If there is anything of value in this book, it goes to my Lord and Savior who has chosen the weak to confound the strong and to my mentor. At the same time, I wish to openly declare that I do not follow men. I do not even follow ideas. I follow the All Knower who is the only source of true truth. I am sure that Dr. Schaeffer as a fellow human being was fraught with human shortcomings. But his example, in spite of that, has been an inspiration to many and most definitely to me.

I will be eternally grateful to Dr. Schaeffer who taught me with love and intelligence that Christianity is not only the best answer, but the only answer. He taught me that the value of a single human being is of infinite worth, as it reflects the very semblance of our loving creator. This is the defining reason that humans stand apart from the rest of creation: They have a true, transcendental significance. And that is the rock upon which all human rights are founded.

My awakening to the crucial need for Christians to be educated and able to respond intelligently to the social issues of our age began in the late 1970s. The issue of abortion had been a low priority in my life until my good friend Max Ault brought to my house the film *Whatever Happened to the Human Race?* created by Franky Schaeffer and his father. Their work completely revolutionized my thinking. And for the first time, I felt that Christians had created a work that displayed intelligence, courage, and sensitivity. It is fair to say that the film opened my eyes to the reality of the eternal cosmic battle between good and evil and to our responsibility to take a stand, no matter the cost.

If we are to be the salt of the earth as commanded by our Savior, then Christians are responsible to know not just *what* but *why* they believe in order to effectively impact their generation. Sadly, most are ill prepared to understand the Judeo-Christian worldview and how it affects the social issues of our day. This book is dedicated to that noble aspiration. It explores the many important social issues of our time and attempts to build a Judeo-Christian response to these vital issues that will deeply affect our children and our children's children.

I am also deeply indebted to Dr. Paul Rodriguez, not only for his steadfast friendship that has never waned through trials and tribulations (a friend who truly sticks closer than a brother), but also for his infinitely valuable suggestions and honest critiques that have helped me bring this work to finality. Iron sharpens iron, and in a very real sense, this is as much a product of his heart as it is mine.

My most pressing motivation to undertake this project is to provide my children a clear and lucid exposition of the infinite value of human life and the need to stand against the tide of our culture, no matter the cost, to defend true truth. It is not our responsibility to win the battle. It is our responsibility to be faithful to the truth and leave the results to God.

It is my most heartfelt desire and prayer that my children will teach their children to teach their children the wonderful gift of the grace of God. This is for you—Michael, Kenneth, Jason, David, Yasha, and Samantha. I am also indebted to my wife, Carmen, for her patience in lending me to the countless hours I have spent in this endeavor and for her steadfast love, support, and faith in me.

This book is dedicated to we the people, the freedom-loving American people, and to those who have sacrificed and continue to work tirelessly in the battle to save the precious little ones whose right to choose life has been callously denied. We may not have victory in our generation, but the final victory has been decided before the foundation of the universe. We stand regardless, leaning on the wisdom and power of the Holy Spirit.

Finally, I am eternally indebted to men of conviction such as Aleksandr Solzhenitsyn who have endured the brutal ravages of political prisons in communist countries because they were unwilling to sacrifice their integrity and principles. They have taught me the meaning of courage and the value of integrity and truth. May God grant us the wisdom and courage to stand firm upon His eternal truths in a world that is becoming ever more hostile to His principles.

In his impassioned plea to the West, Solzhenitsyn warned us of an unfathomable crevasse that threatens to swallow the entire Western world, a crevasse created by a paradigm shift in the thinking of our modern civilization. It is a truly cosmic crevasse that, if left unchecked, will result in a new world order that will infringe on our individual rights in a way that few Westerners can even fathom.

America has now plunged headlong into that crevasse. And the liberty that we have enjoyed in this great country is now in grave danger of being swallowed up into this dark abyss. Unless we understand the true cosmic nature of this crevasse and equip ourselves to intelligently and courageously expose its lethal nature, we will become the victims of our own apathy.

The problem of abortion in America is but a symptom of this great crevasse, but it is not its only lethal symptom. Progressivism is spreading through our youth by the monolithic paradigm that rules our educational institutions. We are losing the future of our nation.

We have myopically viewed all the problems in regard to society and government as disparate bits and pieces of the whole. We have failed to recognize that these separate issues are all symptoms of a fundamental underlying change in the way our society thinks. Our postmodern culture has moved from the Judeo-Christian worldview to the uncritical acceptance of a naturalistic worldview.

This book is dedicated to the task of defending the Judeo-Christian worldview and exposing the great danger of abandoning this fundamental position as our cultural foundation. The social issues that confront us are all related to this paradigm shift in our thinking as a people.

Our fundamental worldview directly affects our stand on social issues. It is impossible to divorce our fundamental beliefs from our laws and ideas of social ethics. This book is designed to help Christians understand the importance of knowing how to give a rational and logical defense of true truth to the watching world lost in the darkness created by this shift.

In his brilliant exposition on our social responsibility as Christians, Schaeffer stated:

> Those who hold the material-energy, chance concept of reality, whether they are Marxist or non-Marxist, not only do not know the truth of the final reality, God, they do not know who Man is. Their concept of Man is what Man is not, just as their concept of the final reality is what final reality is not. Since their concept of Man is mistaken, their concept of society and of law is mistaken, and they have no sufficient base for either society or law.
>
> They have reduced man to even less than his natural finiteness by seeing him only as a complex arrangement of molecules, made complex by blind chance. Instead of seeing him as something great who is significant even in his sinning, they see Man in his essence only as an intrinsically competitive animal, that has no other operating principle than natural selection brought about by the strongest, the fittest, ending on top. And they see Man as acting in this way both individually and collectively as society.[1]

The church at large has failed miserably to prepare our children, much less to win the lost in our generation. We have failed because we have not prepared ourselves to speak intelligently to postmodern man.

We have failed to understand the paradigm shift in a worldview that now stands unchallenged in our higher institutions of education and government. We have failed to expose the wickedness of relativism and the violence it will bring upon the unsuspecting masses who ignorantly embrace it. We have failed to inoculate our children to the seduction of postmodernism and the venom of progressivism.

We have failed to become all things to all people so we might win the more, as the Apostle Paul admonished us. We have failed to love our neighbor enough to learn how to communicate the gospel in the context of this paradigm shift in our culture.

We have failed because pastors have been more concerned about building their own kingdoms than building the kingdom of Christ. We have failed because we are more concerned about filling our pews than truly equipping the saints to be the salt of the earth. We have failed because we are afraid to take a stand on issues that might be controversial—God forbid that our parishioners should become offended and our collection plates suffer. We have simply failed miserably! May God open our eyes and hearts and take away the scales over our eyes that our egos produce. Humble us, Father, and bring us to Your feet at the cross for wisdom to reach our generation in this the highest ever potential harvest of human history. In Your boundless grace we trust.

I am but an ant in the midst of elephants with simply a hope and a faith that truth matters, no matter the cost of speaking it boldly and loudly. I have no illusions about the battle that lies before us. It may happen that within our generation the battle will be lost. Entering into the ancient cosmic battle between good and evil always has consequences. If we dare to be on the front lines, then we must be prepared for the eventual wounds that will surely come. It is the price of battle.

But I rest on this: The war has already been won. We will be victors when our Champion comes. The powers of darkness are great, but they have no power over the light of truth. I have no temporal expectations. My duty is to speak the truth God teaches me and leave the results to Him. The Word of God never returns void.

I am no expert on anything except the knowledge that, as the Apostle Paul also confessed, I am the chiefest of sinners. But I take heart in the truth that God loves to use the weak things of this world to confound the strong. He loves to use the humble to uproot the proud. He loves to use the Davids to defeat the Goliaths. I take courage in my Savior's words:

> *And ye shall be brought before governors and kings for my sake, for a testimony against them and the Gentiles. But when they deliver you up, take no thought how or what ye shall speak: for it shall be given you in that same hour what ye shall speak.*
>
> —Matt. 10:18–19 KJV

Dr. Francis Schaeffer and the author in Ft. Lauderdale, Florida, 1983

Notes

1. Francis Schaeffer, *A Christian Manifesto* (Westchester, IL: Cross-way Books, 1982), 25–26.

INTRODUCTION

● ● ●

WE HAVE FORGOTTEN

My heart breaks for my nation. I try in vain to hold my tongue, but His words are like a fire in my bones. A great and mighty people have strayed from God's path and are heading for destruction. I plead with you to hear the burden that consumes my heart and mind.

Hear this, America. No Gentile nation in the entire history of humanity has ever been blessed with so much abundance. No nation in the entire world has ever enjoyed personal liberty, personal peace, and prosperity as America has. We cried out to God in our hour of need, and He helped us against all odds to defeat the most powerful nation in the world and gain our independence.

We entrusted our future to His providence. He honored our faith and blessed our nation. But, as with every other human institution, we have forgotten who gave us the bounty. We have turned our backs on Him, and truth and justice are now being trampled in the streets.

1

*When I fed them, they were satisfied; when they were
satisfied, they became proud; then they forgot me.*

—Hosea 13:6 NIV

We offer empty rituals and traditions, but not our hearts. In God
we no longer trust. Personal peace at any price has become our idol.
God does not delight in our empty rituals. He wants our hearts.

*For I desire mercy, not sacrifice, and acknowledgment of
God rather than burnt offerings.*

—Hosea 6:6 NIV

Our leaders are brimming with pride. Their words are dripping
with insolence. They are like those who move the boundary lines
and think God is not watching. They cry out to God for public
consumption, empty ritual, and meaningless rhetoric, but their
hearts are far from Him. Their actions betray the real vanity they
treasure in their hearts. They have called evil good and good evil.
Greed has grown deep roots, and our once rock-solid foundation
is crumbling into sand. Do you think our indolence toward truth
escapes the eyes of God? Every nation birthed in truth, even Israel,
has reached this crucial turning point.

*Put the trumpet to your lips! An eagle is over the house of
the Lord because the people have broken my covenant and
rebelled against my law. Israel cries out to me, "Our God
we acknowledge you!" But Israel has rejected what is good;
an enemy will pursue him. They set up kings without my
consent; they choose princes without my approval.*

—Hosea 8:1–4 NIV

The Great Eagle circles above us and sees all. His heart is broken
because we have abandoned His way. Greed and selfishness have
created a spider's web around our people. They set up rulers without

God's consent and bless rulers who answer only to their greed for power and wealth.

> *Woe to them, because they have strayed from me! Destruction to them, because they have rebelled against me! I long to redeem them but they speak about me falsely. They do not cry out to me from their hearts.*
>
> —Hosea 7:13–14 NIV

Our leaders serve for dishonest gain. They become wealthy through the schemes of merchants. But gold and silver shall not save them from God's justice.

> *The merchant uses dishonest scales and loves to defraud. . . . "I am very rich; I have become wealthy. With all my wealth they will not find in me any iniquity or sin."*
>
> —Hosea 12:7–8 NIV

The Lord has a charge to bring to us, America. He will repay us according to our deeds. Our people perish because of lack of knowledge. We look upon the outward appearance, but God looks upon the heart. We do not care for justice. We have been bewitched by the siren song of selfishness and lasciviousness. We are blindly crashing upon the rocks. No one wants to hear this truth. But we shall reap what we have sown.

> *Their treasures of silver will be taken over by briers, and thorns will overrun their tents. The days of punishment are coming, the days of reckoning are at hand.*
>
> —Hosea 9:6–7 NIV

He longs to redeem us. Return to Him. There is yet time. But no matter the outcome, I stand with Joshua at the edge of the River Jordan until the waters part.

Though the fig tree does not bud and there are no grapes on the vines, though the olive crop fails and the fields produce no food, though there are no sheep in the pen and no cattle in the stalls, yet will I rejoice in the LORD, I will be joyful in God my Savior.

—Hab. 3:17–18 NIV

Evil will always seek to find a foothold in this world. We have but to nod in that direction, and it will camp upon our hearts and drive the spike of death deep into our souls. There is no place of refuge aside from death to keep us from the fiery darts of the Great Impostor. But we have a power greater than his villany: the grace of our great God.

He is our shield and buckler and a fortress in the time of need. We have but to humble ourselves to be lifted above the temporal fray. The Enemy of Man can kill our bodies, but he cannot touch our souls. We shall stand next to the Morning Star on the day of the Great White Throne judgment. In that day, at long last, justice shall be executed at the consummation of the ages. The blood of Abel shall be vindicated. We shall overcome.

Thy will be done.

CHAPTER 1

● ● ●

THE GREAT AMERICAN DECEPTION

In 1914, at the beginning of our ill-fated twentieth century, a storm broke over this civilization, a storm the size and range of which no one at that time could grasp. For four years, Europe destroyed herself as never before, and in 1917, a crevasse opened up on the very edge of Europe, a yawning gap enticing the world into an abyss. As Aleksandr Solzhenitsyn explains in *Warning to the West*,

> The causes for this crevasse are not hard to find: it was the logical result of doctrines that had been bandied about in Europe for ages and had enjoyed considerable success. But this crevasse has something cosmic about it, too, in its unplumbed, unsuspected depths, in its unimaginable capacity for growing wider and wider and swallowing up more and more people.[1]

W e no longer stand upon the edge of this unplumbed precipice. We have fallen headlong, deep into the Dragon's ravenous belly. This cosmic crevasse, which was conceived during the havoc and bloodshed of both global conflicts in our last century, did not magically materialize from a vacuum in 1914. As Aleksandr Solzhenitsyn pointed out, this was simply the date that the underlying philosophical ideology grew political teeth.

Neither has it disappeared since the end of the two world wars. It has not even faded since the collapse of the Soviet Union. It is alive and growing like a super-massive black hole that devours with insatiable hunger all who come close to its event horizon. It is a powerful and dangerous spinning vortex that sucks all freedoms and the dignity of the human soul into a dark and deep oblivion that crushes all into indistinguishable sameness. It is the death knell of individuality and personal liberty. It is the monster of an oppressive autocratic centralized bureaucracy that swallows all individual rights and becomes a tyrannical god to all its hapless citizens. This is the natural terminus of socialism.

Not too long ago, America was the bulwark that stood against this cosmic monster spawned in the very bowels of hell. When I fled from the communist island of Cuba, America was the last bastion of great hope for our world. My family lost all that we owned, all that my forefathers for several generations had earned through hard work and diligence. But more painful yet was the loss of my father. I was seven years old when my father was forcefully taken from us and thrown into a political prison at the age of 35—an ordeal that cannot be adequately described with mere human words. Neither can the reader grasp the true measure of pain that my family was forced to suffer.

At the tender age of seven, I felt the sharp talons of this ideology as it yanked from my palms the little paradise I knew and loved, thrusting my family into a maelstrom. In a single moment, everything can be taken from you. Everything your greatgrandfather left for his son and everything his son left for your father can be taken in a single heartbeat. The cumulative effort of many generations can go up in smoke when the ideology of the great crevasse has taken root and flowered.

My father's name was Paco. He had made my grandfather Henry promise that if anything happened to him, he would take my mother and his six children to America where they would be safe. At first, my mother refused to leave my father in Cuba, languishing alone in a political prison. But in time, she realized that she had to place her desire to remain close to her husband aside for the safety of her children. She sacrificed her needs to provide for ours. This is the definition of true, selfless love.

On June 30, 1961, we boarded a Pan American plane to America. The Cuban government had taken everything we owned—our land, our homes, our cars, our savings, our jewelry, our toys, and even my mother's wedding ring. She cried when they took it. And we left with only the clothes on our backs. "It doesn't matter," said my grandfather. "We are flying to freedom, to the land of the free and the home of the brave."

I did not fully comprehend what he meant, but as I grew in knowledge, I came to realize that America was the magical land where anyone who was willing to work hard could make their dreams come true. I became an American citizen in 1976, the bicentennial year, in the city of Philadelphia in front of the Liberty Bell during the Fourth of July celebration. I am an American citizen by choice, not by natural happenstance. I consciously and purposefully joined the land of the free and the home of the brave because there is no other nation in the world like her. It was one of the proudest moments of my life.

But today, only four decades later, I am deeply worried. Will our children be able to enjoy the same freedoms? Will America survive as a sovereign nation? Will our economy survive another carefully orchestrated assault? Will we be forced to coalesce financially with the globalist agenda? Will we rise up and stop our federal government from nationalizing private industry? Will we understand the cost of socialism? Will we understand the consequences of mortgaging our children's future with the astronomical national debt acquired through the brazen implementation of this socialist agenda for our nation? Will they come and take our guns? America has been deceived and lured into the great crevasse. It has been lured because it has not tasted the fruit of collectivism.

The fact that in the 2016 election cycle a candidate for president, Bernie Sanders, who is an avowed socialist and atheist, could almost win the Democratic Party nomination shows how far our nation has fallen. It shows how deeply deceived our young people are. And the fault is ours. We did not prepare them for the progressivist indoctrination that has undermined our public educational system from kindergarten to the highest level of university education.

I am deeply disturbed by the cosmic slumber that has infected the home of the brave. The land will soon not be free. It is as if a magical spell has enchanted the minds of Americans into an apathetic slumber. Our freedoms are being eroded at such an alarming pace that I can scarcely believe my own eyes and ears. Is there courage yet to stem the tide?

Long-cherished ideals that have formed the backbone of our national identity are being discarded overnight. The deep disdain for the centralization of power, which was dearly held by our Founding Fathers and for most of our nation's history, has been forgotten. Instead, a monstrous bureaucracy threatens to make every citizen a slave who must labor for the government for six to seven months of the year before earning a single penny for his or her own family. I dare say that the serfs in the Middle Ages were allowed to keep more of the fruits of their labor than the citizens of modern socialist governments. We in America are well on our way there.

The Constitution is being regarded as a set of anachronistic ideals that need to be reformed in order to meet the "advancements" of postmodern man. Things no longer take years to change as they did in the past. The diabolical blitzkrieg is so quick and the applications so consecutive that it is nearly impossible to mount any concerted resistance.

The monolithic control of the mass media simply serves a globalist agenda by slanting the news to promote a not-so-hidden ideology. The only national media outlets that have not been swallowed up by the left are certain radio talk shows and the Internet. But if I were a betting man, I would put money on it that their days are numbered.

In a single day, financial institutions can crash through artificial manipulations that could bring common people to their knees financially. Cars are repossessed, home loans default, jobs are lost, pensions go bankrupt, inflation turns the dollar into worthless paper, and the public runs blindly to the pied piper with the prettiest flute that will promise to take care of them.

It might go something like this:

"Is that not the most beautiful song you have ever heard?" said the masses.

"Yes, it is certainly pleasing to the ear," said the concerned citizen. "But why do you run off the cliff?"

"Nothing will happen to us. The song will take care of us."

"But the fall is far, and the rocks below are sharp and jagged."

"Why do you always think negatively? Can't you see that everyone is going in the same direction? The whole world can't be wrong, can they?"

"But what about the jagged rocks?"

"You are an obstinate, anachronistic old fool. Get with it, man! We need change."

Politicians with smooth words have lulled the masses into a hypnotized trance and allowed globalists to rob us of our national distinctions. Their goal is to centralize power, diminish the financial and political power of the middle class, and move us toward a global collectivist government that has gone into hyperdrive. Their plan is no longer done in secret meetings but in the open for the entire world to see and to marvel at their cunning schemes.

They have mocked the spirit of our Founding Fathers and the rich heritage hard won long ago by the blood of many citizens since our Independence Day. It is a truth worthy of remembering that most of us do not appreciate our freedoms until we have lost them.

I came from a country where we lost our liberty; it was squandered by petty, greedy politicians on the take. They made us ripe for revolution, and we went from bad to worse. We went from a corrupt dictatorship to an oppressive collectivist form of government that was 10 times more repressive. It was the common people who paid the price in blood and suffering.

I do not take our freedoms lightly. Hear my desperate warning: We are being politically and financially positioned for the final move. The time for the final checkmate is close at hand. Our nation has been maneuvered through clever smokescreens into a deep and cavernous crevasse of cosmic proportions. We have no time to waste.

"What final move?" you may ask.

It is the move to turn us into a socialist nation, which essentially destroys the middle class and their opportunity to reap the benefits of honest labor. It is the move that will rob us of the American dream to leave our children a better life than ours. It will turn the American dream into the American nightmare. It is the move that will make our government capable of uniting with others in a globalist new world order. It is being made possible by the adoption of a relativistic worldview that has overshadowed our Judeo-Christian heritage. It is happening because Christians and Jews have not countered the progressivist, Hegelian worldview.

"Why would anyone wish to do this to America? Surely you are being an alarmist," you say.

There are several vital motives for this continual drift toward a socialist form of government:

1. Socialism is the best way to fleece the masses with massive bureaucracies, which create even more massive national debt. International bankers then finance these national debts and profit quite handsomely at our expense. Imagine what your bill would be if your personal credit card company had the authority to determine how much and when you purchased goods with your credit card and yet the responsibility to pay for it remained yours.

2. It is the most convenient way to merge all the nations into a global government that would allow them to effectively fleece the entire world. The Hebrew Tanakh and the New Testament have long predicted the rise of Babylon, the city that will economically rule the world. In Revelation 18:9–10, we see the global kings and global merchants wailing in lament, weeping and mourning over Babylon's destruction, for they were intricately tied into the global power structure that enslaved all of humanity and allowed them to live deliciously with Babylon.

3. Socialist governments only redistribute the wealth of those who are not the elite in power.

4. Socialist governments are able to harness the support of the much larger element in humanity that thinks they can get something for nothing. Little do they know that in the end, they will pay a great price for their complicity and ignorance.

5. Socialist governments are able to elevate those who are at the very bottom of the economic rung just far enough to keep them from reaching the point of desperation that usually motivates armed resistance. Thus, they give them enough not to starve. But they do not give them enough finances to mount any economic resistance to the elite in power.

6. Socialist governments strip the middle class of the finances necessary to mount any real resistance to the elite in power.

That provides the most efficient way for the elite to rule and fleece the masses with the least overhead costs. It does not create a classless society. It creates two new classes: the rulers and the ruled. Nothing has changed from the tyrannical monarchies and ruling oligarchies so despised by freedom-loving people.

7. Socialist governments are built on relativistic values that allow the elite the freedom to rule without restrictions from individual human rights and moral inhibitions. The individual is sacrificed for the collective.

How Has It Come to This?

The move began long ago when the large monopolies began to push the mom-and-pop stores and family farmers out of business, thus gradually consolidating economic power with an elite few. All the small banks have been gobbled up by a few that rule Wall Street with an iron fist. It will reach critical mass when our government completes the task of nationalizing private businesses, healthcare, food production, transportation, energy, the national media, and the educational systems. It is almost there.

When these things come to pass, we will soon see the loss of our religious freedoms. It will end with the establishment of an international charter that will rule all nations. Our Bill of Rights will be subservient to this international charter.

We are not quite there yet, but we will get there much more quickly than most people think. They have laid the groundwork well, so when the time comes, things will move at the speed of light. A simple, economic, artfully manipulated meltdown of the stock market could easily trigger the final checkmate, not only nationally but also globally.

How has it come to this? How has this cosmic crevasse entered America? It has entered through the clever use of words that are nothing more than clever smokescreens that mask the true horror of the cosmic crevasse. Will we learn from the previous mistakes in our human history? Will we blindly career off the cliff with the herd, following the clever and seductive music of the pied piper?

The pen is truly mightier than the sword, and more often than commonly realized, it has been the pen that has brought the sword to bear upon the flesh of humankind. Through the use of well contrived words, nations have been conquered and hearts and lives have been won and lost. The power of words cannot be underestimated. Words can incite hearts to love or hate and can be used for the greater good of humankind or for unimaginable wickedness and horror.

The Clever Masks of Evil

How can humans be so cruel to other human beings? I have pondered this enigma and have come to understand that the horrendous acts of cruelty done by tyrants and despots always begin with the dehumanization process. They select some class or subsector of our human family and begin to spread propaganda that they are subhuman. When enough people are convinced that these unfortunate members of our family are subhuman, then all violence against them can be rationalized. Hatred has a power to grow exponentially like a fire fed with oxygen once the dehumanization process takes root. Divide and conquer has ever been the greatest tool of the Enemy of Man, and deceit has been next.

The ability to use words cleverly to deceive people's minds has ever been the work of evil despots and tyrants. Throughout the ages, words have been used to cleverly disguise atrocities committed against humankind. They are used as clever smokescreens to mislead the uninformed from the true intents and actions being perpetrated. In most cases, this art of misdirection has proved quite successful. In *1984*, George Orwell introduced the words *doublethink* and *newspeak*, which, when combined to *doublespeak*, describes it succinctly.

The effectiveness of the use of propagandistic euphemisms to mask the real horror of the actions, which they rationalize, has made it an indispensable part of the agenda of those who wish to manipulate our civilization toward their hidden and perverse agendas. They have consistently and effectively succeeded in deceiving the general public through the political art of verbal misinformation and false images. We see these tactics used by both collectivists (communists, socialists, progressivists) as well as supernationalists (fascists, jihadists).

Who can forget the sight of 20 Christian men on their knees on a Libyan beach, each with an ISIS soldier behind them cruelly cutting off their heads and yelling *Allahu Akbar*; or the Jordanian pilot held in a steel cage, doused with gasoline, and burned alive; or the hundreds of Christians crucified on the streets of Syria and Iraq. Such inhumanity always begins with the dehumanization of the victims.

The history of humanity is replete with examples too numerous to recount but none more sinister than the Nazi use of euphemisms to mask the horror behind their diabolical schemes. Hitler was a master of deceptive language, and he used this clever ruse to disguise the atrocities perpetrated against humanity by his regime. German physicians at the Charitable Foundation for Curative and Institutional Care, for example, selected people who were considered terminally ill and euthanized them by gas, medication, or starvation. These "patients" were brought unaware to these killing centers specifically designed for their mass extermination.

Most of us are appalled at the blatant and wholesale murder of millions of human beings that resulted from the elitist mentality that was the natural outcome of the Nazi worldview. We also instinctively reject the inhumanity of those who, in the name of science, used human beings as guinea pigs for pseudomedical or military experiments.

But the Nazis were not alone! The Japanese, who were also supernationalists, established a medical research facility in Manchuria called Unit 731 where experiments such as vivisections were carried out without anesthesia on Chinese peasants. In 1936, by the Imperial Order of Hirohito, thousands of the most brilliant Japanese scientists were sent to this remote area of Manchuria on the outskirts of the city of Harbin to develop bioweapons for the war effort. But that effort was more nefarious than it sounded.

Unit 731 was given the cover of the Epidemic Prevention and Water Purifying Department. The immense compound included a prison that housed some 500 inmates who served as human guinea pigs for the wicked, pseudomedical experiments.

The Japanese secret police rounded up any suspected individuals who voiced resistance to Japanese rule and thus kept a continuous supply of humans for the ghastly experiments at Unit 731. In the scientists' pursuit of biological weapons, they conducted tests on human beings as though they were nothing more than laboratory rats. Sometimes, they killed individuals for the simple purpose of learning their reactions to the various methods used to end their lives.

Their calloused disregard for the value of human life was the natural outworking of a religious worldview that was pagan at its core, worshipping Hirohito, a human, as their god. There is no place for absolute morals under such a philosophical system. Whatever Hirohito said was law. Hitler, Hirohito, or Islamic autocrats—there is no difference. They pervert the true form of nationalism and profane it into a supernationalist, elitist worldview that dehumanizes all others as inferior and dispensable.

All choices are essentially relative outside of a singular, personal creator God who stands above all humankind and even outside our universe and from whom ethical absolutes are derived. Morality becomes what is expedient for the powerful elite. And they justify their acts of barbarism by their political objectives. Pragmatism devoid of any ethical concerns becomes their god.

The use of clever smokescreens to mask the horror of their deeds was just as prevalent with the Japanese as it was with Nazi Germany. The Japanese scientists referred to their victims euphemistically as *marutas*, or *logs*, to fuel their incinerators that burned day and night. This semantic ploy was used in order to somehow evade in their conscious minds the humanity of their victims. These are the universal tactics—the clever smokescreens—used by those who rationalize their cruel brutality to assuage their consciences and mask their despicable crimes.

The Japanese scientists' insensitivity to the dignity of human life was underscored by their calloused disregard for the suffering of their patients as they carried out their heinous experiments without anesthesia. These so-called scientists studied the effects of poisons on their hapless victims and surgically carved open their bodies to see the physiological impact of the poisons on the organs, all while the patients were still alive.

Men, women, and children of every age were intentionally killed by every conceivable method, from the use of poisons and acids to freezing them slowly in order to study the effects of frostbite. Thousands were injected with diseases such as the Plague, cholera, typhoid, defoliation bacilli, and gangrene in order to research potential bioweapons.

Some were given medicines to determine the effectiveness of antidotes, while others were operated on during the various stages of the development of the disease. The scientists thus learned the exact damage to various organs of the body during the stages of a disease.

These experiments were not carried out to provide medical assistance; they were carried out to create more effective weapons of war. Experiments with explosives were also carried out on humans as if they were lifeless mannequins. Human beings were tied to planks in concentric circles spanning out from the center where an explosive device was detonated. The humans were tied to the wooden platform so they stood erect. After the device was detonated, the scientists meticulously studied the shrapnel patterns on the bodies.

But do not imagine for one moment that surviving the blast would somehow give them a reprieve from this nightmare. Those in the farthest circles who were unlucky enough to survive the blast were later killed in some other equally ghastly experiment.

***

No one who ever entered the holding cells of Unit 731 ever came out alive—ever. The ovens burned day and night, incinerating thousands of humans after they had been inflicted by untold barbaric acts. The diabolically clever and sinister use of the word *marutas* (burning logs) was specifically designed and intended to dehumanize and depersonalize the victims in order to superficially assuage the guilt of the scientists.

***

Throughout all time, the architects of deceptive euphemisms have attempted to subliminally mask the obvious horror of calloused crimes. But these architects will not escape the scrutiny of the final judgment. Ultimately, they will all be held liable for the consequences of their brutality because there is One in heaven who will bring every human intention and action into account and one day universally serve justice on all perpetrators of such evil.

The voices of the innocent cry out for justice from the blood-soaked ground of our wilting planet. It has done so from the time of Abel, and you can be sure that no human or superhuman agencies will ever be able to conceal from the Holy One their wicked deeds, no matter how covertly cloaked they are in the darkness created by their smokescreens. Justice shall one day be concluded. All inequities shall be brought to light and remanded. Of this you can be sure: We will all reap what we sow.

The insolent and impenitent voice of Cain callously echoes throughout all generations: "Am I my brother's keeper?" (Gen. 4:9). With hardened hearts and insensitive disregard for the sanctity of human life, the Cains of this world defiantly brutalize their brothers and sisters for selfish gain. And so, perfectly aligned with this demonically inspired and ancient human sentiment first expressed by Cain, the very first human born on our planet, the Japanese camouflaged their true intent behind the subterfuge of seemingly benign walls, cleverly disguising it as a water purification plant.

And there, behind those human-made walls, they artfully concealed the true horror of Unit 731, brazenly brutalizing their Chinese brothers and sisters in seeming impunity. Spurred by their myopic view of reality that left them unfettered by any moral constraints, they incessantly pursued their perverse search for greater military prowess with their unbridled disregard for the sanctity of human life. They callously indulged in numerous grotesque tortures for the sake of advancing what they called scientific knowledge.

Scientific knowledge, which is not evil in and of itself, was used as a weapon, like a gun. In the hands of selfish people, that weapon can be used for great evil. In the hands of selfless people, it can be used for great good.

But when it is made into an idol whose veneration is rewarded by real worldly power, when it becomes the center of things from which all things are measured, it becomes a god. And then it becomes the greatest and most powerful means to achieve the wicked schemes of the enemy of God. The addicting and intoxicating mirage of power has led not a few down this beastly and bloody trail.

The brutality of Unit 731 was not limited to their ghoulish laboratory hidden behind the deceptive walls. The cloak of these criminals' anonymity naturally bred a false sense of security that resulted in a heightened, brazen arrogance. And so, spurred by their initial success in their secret confines, they moved out of these walls into the community at large.

Clay jars filled with fleas that carried the Bubonic Plague virus were tied to balloons and dropped on unsuspecting Chinese villages to field-test the rate of spread and lethal effectiveness of this particular biological weapon. As a result of their field tests, the Bubonic Plague decimated the populations of many unsuspecting Chinese villages from 1939 to 1942.

In some villages, more than a third of their inhabitants succumbed to the horrors and great suffering of this so-called Black Death. Scores of Japanese scientists, dressed in white isolation gowns and self-contained breathing apparatuses, descended on the fields where Chinese people were working and performed open-air vivisections to study the effectiveness of their field weapons.

On other occasions, they gave candy laced with anthrax spores to innocent Chinese children. This was designed to study the effectiveness of a rudimentary, inexpensive means of infecting an enemy with ghastly diseases in order to either incapacitate or annihilate their youth.

But the Chinese were not the only ones to die because of these scientific experiments. Ironically, their records show that more than 1,600 Japanese died as a result of accidental contamination.

Sadly, when World War II ended, there were 400 inmates, still alive in their holding cells, who could have been set free. But every one of them was summarily executed in order to keep the ghastly experiments secret. And yet even more disturbing was the fact that these barbarous deeds perpetrated on innocent human beings under the guise of scientific experiments went completely unpunished by the war tribunals.

Apparently, the supposed scientific value of the volumes of meticulous notes on the countless experiments was used as a bargaining chip to buy unconditional immunity from prosecution for the perpetrators of these heinous war crimes. The American government, eager to get firsthand, scientific, medical information about what was carried out on live human beings, covered up what actually happened and granted unconditional immunity in exchange for the scientific information contained in the records of the nightmare of Unit 731. Needless to say, the American moral compass strayed far from its true north.

The Voice of Abel

With great sadness, we must acknowledge when all is said and done that America's hands were equally stained with the blood of these innocents so callously brutalized for "scientific" purposes. Our hands were stained with the same blood of Abel that cries from the ground for justice, even to this day. America stood hand-in-hand with Cain and callously exclaimed, "Am I my brother's keeper?" (Gen. 4:9). How easy it is to justify guilt!

So it is clear for the objective mind to see that these historical examples of unfettered brutality against humanity demonstrate that there was no difference in the atrocities of the Nazis and the atrocities of the Japanese. Nor are the Americans whose rationalizations allowed these so-called scientists to literally get away with murder less culpable. Every nation at one time or another has walked down the same dark corridors of greed, lusting for scientific knowledge that can be used to attain world power.

This was not America's first forage into this dark cavern. American scientists had already been tainted with innocent blood decades before this sad chapter in our nation's history. The voice of Cain had already bellowed clearly for some time in certain circles of our scientific community.

Prior to World War II, in the 1930s, American scientists involved in a research project called the Tuskegee Study of Untreated Syphilis in the Negro Male intentionally did not advise some 200 black men that they had contracted syphilis. They purposefully kept the men ignorant of the disease in order to scientifically study its long-term effects on living human beings.

In other words, they intentionally withheld medical services that could have alleviated their condition in order to have live human guinea pigs to study the ravages of the disease throughout its many stages. These so-called scientists intentionally and callously watched as this deadly, debilitating disease slowly ravaged these 200 black men who were sacrificed on the altar of science. How does humanity stoop to this level?

How can human beings rationalize such cold, calculated brutality against innocent human beings? Is this wholesale disregard for the dignity and value of human life wrong? Is it really evil to kill another human being? Can we condemn the actions of such people as immoral?

Is it immoral to sacrifice the health of some subsector of our society in order to benefit the health of others? Were these American scientists involved in the Tuskegee study so convinced of the subhuman status of black men that they could rationalize such barbarity against them? Do we have a right to victimize the lower economic sectors of our society in order to benefit the more privileged?

***

Only if we understand the philosophical foundation of those who have managed to rationalize such dehumanization will we be able to understand their ability to inflict such cruelty upon other human beings. But more importantly, only then will we be able to prepare ourselves in order to attempt to avert such barbarism in our future culture.

***

The nature of our philosophical foundation produces very real and tangible differences in the type of choices they create in the scientific and political arena. And in turn, these choices invariably have real consequences on our culture at large and are therefore directly related to this initial philosophical framework.

Good choices lead to good actions that result in good consequences. Bad choices result in bad actions that invariably lead to bad consequences. But our choices are determined by our underlying worldview. The adoption of a relativistic worldview erases the boundaries between good and evil and is simply replaced by pragmatism. Hence, our philosophical understanding of the universe and the nature of humankind deeply affects the consequences we experience individually and collectively as a nation. For this reason, understanding the true nature of the relationship of humans to God is a categorical imperative for all cultures. Knowing right and wrong matters. Truth matters.

There is in this world an inevitable and consequential cause-and-effect relation between the fundamental worldview adopted by the culture at large and the reality it creates in every aspect of the human endeavor. It is an inescapable chain reaction that is absolutely unavoidable. Our foundational philosophical framework provides us with a predisposition for a set of choices that, in turn, results in action. In the end, these actions result in real consequences that are therefore directly linked and flow from the underlying worldview.

In other words, what we believe directly leads us to a specific set of choices, which subsequently leads us to specific actions that impact every aspect of our society. And these actions then produce specific consequences that are directly linked to the foundational philosophical framework from whence they initially developed.

The modern dualism that seeks to separate science from our metaphysical choices is nothing less than another smokescreen to legitimize the teaching of relativism through naturalism in our educational systems. Our public educational system is, in essence, an indoctrinational system that creates in the minds of our youth the impression that science supports an atheistic and relativistic worldview. No other worldview is permitted to present scientific evidence to the contrary. Eminent scientists who are every bit as qualified as any evolutionist are banned from expressing their scientific findings simply because they do not hold to the evolutionist's metaphysical position. It is in every respect a repressive system that is jealously guarded by the priests of scientism.

The impact of this modern paradigm is deeply affecting the future of our nation. Western culture no longer holds to the high view that humans are created in the image of God. It is no great secret that our scientific and political elites jumped across a great divide in the early part of the twentieth century. This is the great divide spoken of so prophetically by Solzhenitsyn. And it has become evident that our general American culture reflects this major paradigm shift today. We have plunged deeply into the dangerous cosmic crevasse that Solzhenitsyn so clearly forewarned us about. We as a culture have abandoned the absolute values of the Judeo-Christian worldview in favor of the relativist values of naturalism. And the bitter fruit of that

THE GREAT AMERICAN DECEPTION

can be seen by the alarming increase in mass murders, even within our nation's schools. It is directly related to the teaching that human life is just the product of an impersonal, chemical accident.

The Hoodwinking of America

The clever smokescreen of useful euphemisms has once again charmed and deceived the masses. Take the word *progress*, for instance. What person could possibly be against progressing or moving forward to a better place? It is a tremendously appealing word, rich with positive meaning and hope. Is not the essence of the American dream that each generation can work hard and progress to enable the next generation to live at a higher standard? Who can be against progress?

But the clever smokescreen created by the term *progressive* is diabolical. There is no progression forward in personal liberties or economic opportunities created by the progressive movement. There is only an increase in the authoritarian rule of the federal government and an increase in the taxation of working people, which impedes them from progressing forward financially. It cements the working person to a stagnant economic level that cannot be surpassed no matter how hard he or she works.

Progressivism is nothing more than a gargantuan siphon that unjustly sucks the money from the pockets of working individuals to redistribute it through an inefficient bureaucratic mechanism that profits handsomely by siphoning its own share of that money to feed the monster of national debt that will eventually completely collapse our economy. The movement at best should be renamed the Stagnatist Movement because the economy of every nation that has practiced it has stagnated. This is a historical fact.

We can point to Cuba, Venezuela, the Soviet Union, China, North Korea, and many other collectivist governments that have brought not only economic ruin but also the loss of personal liberty to the masses. It could more accurately be labeled the Regressivist Movement since it sacrifices our hallowed individual freedoms for the so-called sake of the collective and returns our political system to the very tyrannical concepts of the centralization of power that our nation so bravely rebelled against during the American Revolution.

At the very core of this progressivist philosophical foundation is the belief that there is no absolute truth or absolute standard of morality as expressed in the traditional Judeo-Christian worldview. We are no longer able to make good choices because all choices have been declared equally valid. Hence, right becomes what either the majority vote dictates or the elite in power force upon us. We have been hoodwinked into thinking that political ideologies have no connection to metaphysical worldviews. All political ideologies are rooted in a metaphysical worldview. The progressivist nomenclature is, at its core, simply a Madison Fifth Avenue propaganda attempt to make more palatable for the masses the collectivist ideology of communism.

Our culture has been socially conditioned to become automatically incensed at those who dare to oppose their enlightened ideology of relativism and the political ideology of progressivism through the false division they make between our metaphysical worldview and our political ideology. The idea that there are no absolute truths is portrayed as rational and progressive thinking. Those who espouse the Judeo-Christian worldview are branded as narrow-minded moralists and are arrogantly looked down upon as ignorant, regressive, and superstitious people who should be barred from competing in the public square.

The rallying cry of the relativist is the separation of church and state. But the meaning they derive from that goes far beyond the church as an institution. They want to eliminate any connection between the Judeo-Christian worldview and the political system. How often have we heard this sentiment expressed in one form or another in our institutions of higher learning and the corridors of political power in our nation's capitol? It is the spirit of our age.

The cultured person is expected to portray a spirit of tolerance that legitimizes all other worldviews as equally valuable. That is nothing more than naked relativism. We certainly believe that everyone has a right to make choices, but there is no guarantee that all choices are right. Reason, not superstition, provides the rational basis of understanding for those who truly want to understand what is right. And only by discerning what is right can we make laws that reflect righteousness and justice.

It is reason that brings us to the conclusion that all crimes against humanity, as we have previously underscored, cannot be condemned or even categorized as evil if there is no God from whom we can determine what is right and wrong. Relativism is not a rational belief but rather an emotional belief blindly accepted by those who do not want to feel guilty about the moral choices they make. It is an escape from reason, as Francis Schaeffer so aptly described.

This is the great smokescreen that shrouds the minds of people trapped in this deep, dark crevasse that Solzhenitsyn warned us about. It is the smokescreen that can effectively mask the wickedness of certain choices. It is the fruit of relativism unrestrained.

To be sure, we are not alone as a nation. Perhaps we are the very last Western nation to have plunged into this dark crevasse. All of Europe certainly has already fallen off the cliff and plunged headlong into that crevasse.

Solzhenitsyn was quite right, for they fell into the crevasse long before America. His beloved Russia fell into it in 1917. Ever since, the crevasse has rapidly grown wider and wider. It has cleverly mutated to take on more appealing veneers such as progressivism and socialism. But they are nevertheless the same sinister Hegelian crevasse.

The sad truth is that this crevasse has crossed the Atlantic and already swallowed up a large swath of our nation. None can deny that today; we in America are living in a post-Judeo-Christian society. Our paramount foundational premise of the inalienable and God-given rights of the individual has been literally obscured by our rising new world order in which individual rights must be sacrificed by the tyranny of the majority.

We have been hoodwinked into accepting the relativistic philosophical and moral substratum that readily justifies all these abuses to humanity. We have been hoodwinked into relinquishing the only concrete structure from which the rights of the individual could be assured.

The naïve notion bandied about that we must separate our religion from our politics is nothing but a sinister smokescreen to mask the real intent. It is impossible to separate one's religious worldview from one's political ideology. Our foundational worldview is wholly inseparable from the type of political choices we make. But because our nation has been accepting the relativistic Hegelian naturalistic paradigm as its foundational worldview, the high view of humans once held by the Judeo-Christian mindset of the past has seriously eroded. Relativism leads to pragmatism and erases the very foundational premise upon which our nation was built. Individual liberty is then inevitably bound to be sacrificed by the tyranny of the majority consensus.

We have been scolded, chided, and intimidated into keeping our religious precepts separate from our governmental policies. Contrary to that insistence, our political philosophy cannot be divorced from our foundational worldview. But if we are careful to read between the lines, what they are really saying is this: I only want my antireligious views to be implemented by our government. Atheism is a metaphysical worldview. Naturalism is equally a metaphysical worldview. Their interconnection cannot be ignored.

Moreover, if we prohibit only Christians and Jews from exercising their rights to reflect their foundational worldview through the democratic process, then we are no better than a repressive tyrannical form of dictatorship such as those we hypocritically condemn in other parts of our wilting planet.

We have moved far from our foundational premise that our inalienable rights are intrinsic to all humans because God has endowed us with them. Our government has not given us these individual rights; God has. If it were the government that granted us these rights, they would not be inalienable. The government could take them away again. This is the fundamental truth, explicitly understood by our Founding Fathers, that fueled the American Revolution. We have strayed far and wide from those heady days. The yawning crevasse is growing each year exponentially and will swallow us completely if we do not understand it and begin to oppose it.

The Articles of Incorporation of the United States of America and Islam

God is the author of our human rights. From the very first document drafted by our Founding Fathers, this absolute truth was recognized as a self-evident reality. The American Declaration of Independence states:

> We hold these truths to be self-evident, that all men are created equal, that they are **endowed by their Creator with certain unalienable Rights,** that among these are Life, Liberty and the pursuit of Happiness.—That to secure these rights, Governments are instituted among Men, **deriving their just powers from the consent of the governed.**—That whenever any Form of Government becomes destructive of these ends, it is the Right of the People to alter or to abolish it, and to institute new Government (emphasis added).[2]

Four fundamental and indispensible truths were understood from the very inception of our nation, which made this form of government the following unique American experiment:

1. That our individual rights are granted to us from the supreme creator of the universe alone
2. That the government has no authority to intervene or hinder these individual human rights because they are unalienable rights

3. That the power of the government is derived only from the consent of the governed
4. That if the government becomes destructive to those ends, it is the right of the people to alter and abolish it

Our government is charged with the duty to secure those individual rights. If our government fails to secure those God-given rights, then it is the right of the people to either alter the government or abolish it. The government is not an authority above the individual and cannot deny those basic human rights. Its power does not come from an authoritarian, centralized power structure but from the very consent of the people whose representatives are elected to govern and protect them. This is the essence of our American experiment that electrified the entire world and set the nations ablaze with the desire to also have these freedoms.

The supremacy of the Declaration of Independence, a foundational document for our nation and its governmental policies, must be clearly understood. It is not only chronologically preeminent but also thematically unsurpassed. The Declaration of Independence is, in fact, the charter of our nation and can be compared to an organization's articles of incorporation. The Constitution is comparable to a corporation's bylaws.

The articles of incorporation is the charter that makes a corporation a legal entity. It defines that entity. It provides the very essence of the purpose and intent of that organization. Bylaws are the rules that govern a corporation's function. So it stands to reason that bylaws should never be interpreted or intentionally twisted in a way that would contradict the purposes laid out in the initial articles of incorporation, which is the express nature and character of that corporation.

It is therefore clear and indisputable that our Founding Fathers saw the Declaration of Independence as the higher document of the two. Moreover, the express declaration in this preeminent and primordial American document is that God has given every human the right to freedom. No more precious word exists in our vocabulary than *freedom*. It is liberty that allows a culture to thrive and really progress. This right to liberty given to us by God includes humans of every color, creed, age, ethnic background, financial status, or any other superficial distinction we can make.

Any judicial interpretations made even by the United States Supreme Court, the highest court in the land, that countermand these foundational premises are illegitimate. Any laws passed by Congress that countermand these basic principles are illegitimate, even if they constitute the will of the majority of the people. Any executive order signed by the president of the United States that countermands the principles in the Declaration of Independence is null and void. Any law instituted by any state that countermands these fundamental principles is automatically invalid. The rights of each and every individual cannot be sacrificed for the benefit of the majority or the elite in power.

Penned by the very hand of Thomas Jefferson and amended by the Continental Congress to include explicit references to God, our foundational charter as a nation stipulates that our individual rights are granted by the great creator and not by any branch of our government. There is no wall of separation between government and God.

As a matter of fact, the Declaration of Independence is a reasonable and rational appeal to the Universal Judge for the rectitude of its claims. It begins with God and ends with the acknowledgment that God is the protector and supreme judge of the world, not just our nation. Furthermore, it stipulates that our nation relies upon and trusts in the supreme creator for protection through His divine providence.

We, therefore, the Representatives of the United States of America, in General Congress, Assembled, **appealing to**

the Supreme Judge of the world for the rectitude of our intentions, do, in the Name, and by Authority of the good People of these Colonies, solemnly publish and declare, That these United Colonies are, and of Right ought to be Free and Independent States. . . . And for the support of this Declaration, **with a firm reliance on the protection of divine Providence**, we mutually pledge to each other our Lives, our Fortunes and our sacred Honor (emphasis added).[3]

Our Founding Fathers declared that it was God who gave us the right to be a nation. They, through reason, expressly stated to the public at large that it was to Him, the supreme judge of the world, that they brought their grievances for the legitimization of their actions. Moreover, they stated that they relied completely on His hand of providence to carry it out and protect the nation to the end. That is the America I signed up for when I became a citizen. It is to that revolutionary idea that I pledged my life, my fortune, and my sacred honor. I am willing to fight and die for that America.

For this reason, we must be quite circumspect in those we permit to immigrate to our nation. If they do not hold the documents that describe the very essence of our national identity as preeminent over all other forms of government, then they should not be allowed to cross our borders. Our nation is multiethnic, and that is a wonderful thing. It matters not what color your skin is, what nationality you come from, what God you worship, or even if you do not worship a god. The requirement to become an American citizen is to adopt the nation's national identity and the fundamental elements that make it the most unique, sovereign nation in the entire world.

But it does matter if your religion is teaching you that your purpose in immigrating is to overturn the present culture of the nation and undermine the legitimacy of its constitutional form of government. That is called sedition, and we must not allow progressive courts to legitimize sedition against our American principles of self-government. It is time to stop this form of legal treason in our nation, or our national identity and our Constitution will be profaned.

The modern secular myth that there was an absolute wall of separation between faith and government at this defining moment in our nation's history is but a surreptitious lie forged in the deepest recesses of hell. It is promoted by the humanist sectors of our society that want to expunge any memory of the legitimate influence the faith of our Founding Fathers had on the design of our government and turn us into a secular, socialist system of government.

The Rise of the New Age

We have strayed far from America's foundation. Today, many modern people in the West have chosen to camouflage this naturalistic crevasse in the Eastern garb of pantheism—not because there is any real movement to escape this crevasse but because of aesthetic reasons.

The sterility of the naturalistic paradigm inevitably leads humans toward a pantheistic compromise that is more aesthetic, at least on a superficial level. It adds a religious dimension to the same naturalistic philosophical foundation shared by both worldviews. Pantheism lays an illusory floor across the gap of the vast crevasse.

But that seemingly sturdy floor spanning the crevasse is just a hologram. The rise of pantheistic mysticism and occultism in our culture cannot fill the void left by the abandonment of the Judeo-Christian worldview. Pantheism and occultism cannot provide a basis for the high view of humankind.

Our individual right to self-government and the freedoms God has given us are the direct consequences of the preeminent founding principle that we are significant transcendent beings created in God's image. That is the exclusive domain of the Judeo-Christian worldview. No other worldview has ever made that claim.

Moreover, the government has no right to take those freedoms from us because the creator, not some bureaucratic power structure endowed us with them. Our political leaders are our servants, not our masters. That is the very heart of the Declaration of Independence and the pillar of our American experiment. Without it, we have no basis for a system of jurisprudence that can defend the individual against the will of the collective or power elite. It is for that reason that the Constitution established the Senate and the House of Representatives (we the people) as the overseers of the federal government.

Pantheism, occultism, and atheism are all equally incapable of providing a substratum from which one can build a coherent framework of morality and justice that could defend the rights of the individual as prescribed by the Declaration of Independence. Under the pantheistic religion accepted by most of Japan during World War II; under the atheistic occult philosophy expressed by most Nazi fanatics; under the secular atheistic worldview of communism, socialism, and progressivism; and under the naturalistic worldview of our modern culture, there is no absolute paragon or singular personal God from which one can derive a coherent and valid standard for morality, truth, justice, or the transcendental value of the individual. Hence, there is no absolute basis for the rights of the individual. Relativism is the natural nemesis of individual rights.

Pantheism can only yield relativism. If God is all, then there can be no absolute delineation of right and wrong. All is sameness. Whichever side of the "force" one chooses to call evil, another can, with the same legitimacy, choose to call it good. They are both sides of the same force.

And it naturally follows that if truth and morals are relative, then we have no right to condemn the previously mentioned brutal and savage actions against fellow humans. We have no grounds upon which to call these actions immoral. If there is no final resting point outside of humans upon which we can rationally pin what is moral and immoral, then all choices are essentially equal.

In this case, the measure of morality would be predicated on the personal choice of each individual, and all choices would thus be equally valid if there were no God above us to declare otherwise. Each and every human's personal moral choice would be intrinsically as valid as the choice of any other. Hitler's morality would be as valid as Mother Teresa's.

But are morals really relative, or is there an absolute basis from which we can ascertain absolute morality? Is truth really relative? Is the idea of justice and personal rights simply an unreachable illusion in our dreams? If there is no God, then why do we even have the notion of good and evil? Why do we yearn for justice when we are faced with great cruelties?

That is the very substance of this cosmic crevasse. And if we are to survive as a nation that has historically championed the ideals of freedom and liberty, we must understand the ever-escalating danger that looms before us. Make no mistake, we are facing a critical juncture in our modern civilization, and our choices today will have eternal consequences on the future of the free world and the freedoms our children will have.

Each and every one of us is therefore equally responsible to rationally and objectively consider the foundational questions of life. This book has been written with the express purpose of providing for the searching, benevolent, human mind a rational and objective analysis of the contemporary social issues of our day in the hope of stimulating our responsibility as human beings.

In spite of the great foundation that birthed our nation, it must be noted that the principles were not perfectly carried out. Our nation was born with a great birth defect. The principle was right, but the execution of that principle was definitely not. The principle is that all humankind was created equal. Therefore, the denial of any God-given rights to any sector of humanity is illegitimate from the Judeo-Christian perspective. Hence, slavery is an abomination in the eyes of those who truly understand the true message of Genesis in the Hebrew Scriptures and the New Testament, revered both by true Jews and Christians. These fundamental truths are inviolable:

1. All humankind has been made living souls by the *nashama* (breath) of God. "Then the LORD **God** formed man of dust from the ground, and **breathed** into his nostrils the breath of life; and man became a living being" (emphasis added) (Gen. 2:7).
2. We are all children of Adam and Eve. We are all from the same parents 10 times over between Adam and Noah.
3. All people are equal in God's eyes. "Then Peter opened his mouth, and said, Of a truth I perceive that God is no respecter of persons: But in every nation he that feareth him, and worketh righteousness, is accepted with him" (Acts 10:34–35 KJV).

Some who call themselves Christians may be in conflict with this third truth. But the truth is absolute, not because Christians believe it but because God has ordained it so. Those Christians and Jews who go against these truths are in rebellion to the Judeo-Christian worldview.

However, the relativist has no such claim. He or she cannot object to another person's choice to enslave other fellow human beings on any moral grounds. Both choices are equally valid under a relativistic ideology. Pragmatism becomes the moral imperative, and if it is profitable to have slaves, then it is not immoral. As a matter of fact, it was Christianity that stood as the bulwark against slavery and eventually brought freedom to our African brothers and sisters in the West.

Notes

1. Aleksandr Solzhenitsyn, *Warning to the West* (New York: Ferrar, Straus and Giroux, 1976), 128.
2. National Archives, "Declaration of Independence," July 4, 1776, https://www.archives.gov/founding-docs/declaration-transcript.
3. Ibid.

CHAPTER 2

● ● ●

ENTERING THE AGE OF VERICIDE

In addition to the grave political situation in the world today, we are also witnessing the emergence of a crisis of unknown nature, one completely new, and entirely non-political. We are approaching a major turning point in world history, in the history of civilization. It has already been noted by specialists in various areas. I could compare it only with the turning from the Middle Ages to the modern era, a shift in our civilization. It is a juncture at which settled concepts suddenly become hazy, lose their precise contours, at which our familiar and commonly used words lose their meaning, become empty shells, and methods which have been reliable for many centuries no longer work. It's the sort of turning point where the hierarchy of values which we have venerated, and which we use to determine what is important to us and what causes our hearts to beat is starting to rock and may collapse.[1]

In this grave hour in our human history when the forces arrayed against our Western civilization have never been as powerful and lethal, when their promoters have entrenched themselves in such high places and are financed with more wealth and technological prowess than ever before, our responsibility to stand in the gap, to resist with love and intelligence the machinations of these evil forces, is more crucial than ever before.

Upon the shoulders of our generation at this crucial juncture during the turn of the third millennium rests the future of the entire free world. A global tyranny looms ever closer on the horizon as nefarious forces weave their economic spider's web to trap all nations. Europe has already succumbed. Some in America have adopted a fatalistic mentality, citing the fact that the scriptures foretell a time in which the world will be ruled by the Antichrist and therefore rationalize doing nothing.

This may be true, but they fail to understand the deep truth that in every generation, we are to stand for the truth and fight the battle of the long defeat no matter what the odds are against us. We are to fight evil no matter the cost and no matter our chances of winning. This is our task in every generation.

Our choice to enter into the fray should not be made by the cold calculations of the probability of being physically victorious. Victory in God's economy is being faithful. We fight regardless of the immediate outcome. We fight to be faithful and leave the results to the great creator. We fight by His power and light and lean not on human strength and understanding. We fight, even when there is no chance of victory from our human perspective. We fight because we know it is the right thing to do, even when no one else fights alongside us. We fight because we know that our great Savior has already won the final victory.

Our hope springs eternal from this divinely ordained fountain, so we place our trust in His providence, and by His grace we stand. Our weapons are not temporal in nature. Our weapons are truth, love, and perseverance borne on the wings of the Holy Spirit. And for this reason, our rewards are consequently also eternal in nature. Our treasures are everlasting and incorruptible. No one can rob us of them. No rust can corrupt them. Moths cannot destroy them. They will not fade with time. They are more secure than the ground beneath your feet. This is our calling in every age.

Will we continue to read books that only entertain us? Sadly, most people today do not read books anymore. They have been programmed to lose patience with an endeavor that requires some time to complete. Will we make the mental effort to understand the philosophical position that has empowered this political attack upon our cherished freedoms? Will we take the time and effort to educate ourselves in order to communicate intelligently to our generation? Will we stand on the front line where the battle is raging? Will we resist the spirit of this world? Or will we hide within the cloisters of our modern monasteries and be content to sing songs to one another?

The Christian is to resist the spirit of the world. But when we say this, we must understand that the world-spirit does not always take the same form. So the Christian must resist the spirit of the world *in whatever form it takes in his own generation.* If he does not do this, he is not resisting the spirit of the world at all. This is especially so for our generation, as the forces at work against us are of such a total nature. It is our generation of Christians more than any other that needs to heed these words attributed to Martin Luther:

> If I profess with the loudest voice and clearest exposition every portion of the truth of God except precisely that little point which the world and the devil are at that moment attacking, I am not confessing Christ, however boldly I may be professing Christ. Where the battle rages, there the loyalty of the soldier is proved, and to be steady on all the battlefield besides, is mere fight and disgrace if he flinches at that point.[2]

Solzhenitsyn's warning more than a quarter century ago has also proved to be quite prophetic. That shift in paradigm that he solemnly predicted has unfortunately arrived in America. Sadly, the lethal consequences it has left in its wake can be eclipsed only by what will happen in the future if we fail to have the intelligence and courage to oppose it today.

How is it that this shift in paradigm has gained such a foothold in our nation founded only two centuries ago on a Judeo-Christian premise that was at its bedrock the exact opposite of this naturalistic paradigm? How is it that most of our postmodern society has come to accept as a fundamental premise the presumption that truth and morals are relative?

It is not because the naturalistic worldview is a better alternative to the Judeo-Christian worldview. It is because Christians and Jews have failed to intelligently oppose this cancerous ideology. It is because we fell asleep at the wheel, and now we are finding ourselves neatly contained and tied up in the trunk while someone else is driving the car.

We must awaken from our magical enchantment and learn to think and act as the bearers of the torch of truth that alone can lead to liberty, justice, and equality for all human beings. This is the duty of every American citizen and specifically for those who hold to the Judeo-Christian worldview. It is the Judeo-Christian worldview that alone forms the basis for liberty, justice, and equality. Let me repeat that another way. No other worldview is capable of establishing a foundation for our individual liberties. If you want to know why, then keep reading.

Our duty to resist the spirit of this world is not new to our generation. This cosmic battle has existed in every generation since Adam. It is our part in the greater battle of the long defeat that has raged since the Garden of Eden to resist the forces of evil from achieving their esoteric plans. But as we near the time of the end, our action—or inaction—impacts more lives than ever before in human history, and more souls are at stake.

For such a time as this, God in His providence has placed you and me. We must educate ourselves and learn to refute the ideology that is destroying our nation and robbing our children of the liberties we have enjoyed. This is our watch. Will we sleep? Will we take the mental effort to learn how this dastardly deception has infected our nation?

The following chapter is not an easy read. It forces the reader to think. You may have to read the sentences more than once to completely grasp their meaning. But can we afford not to make this small sacrifice in order to learn what tools we need to confront the cosmic crevasse that is engulfing our future?

The Hegelian Dialectic

The Age of Vericide began with the advent of Hegel's dialectic as the Darwinian paradigm of natural evolution was applied to the realm of philosophy. Up to that point in human history, humans thought in terms of antithesis (opposites), as Francis Schaeffer illustrated so well in his extremely important trilogy—*The God Who Is There, Escape from Reason,* and *He Is There and He Is Not Silent*—must-reads for any thinking person. In other words, our civilization was founded on the accepted idea that truth was absolute. Common logic dictated that if a certain thing is true, then the opposite is false.

But today, modern thinkers have abandoned this concept as anachronistic and accepted the naturalistic paradigm that denies the existence of absolute truth. They have accepted the idea that truth is relative; to be more specific, that truth evolves.

Georg Wilhelm Friedrich Hegel proposed that truth was ever-changing and evolving. He theorized that truth was not absolute and that in time, it evolved through the process of synthesizing. Through Hegel's dialectic, the traditional and logical process in the determination of absolute truth (thesis/antithesis) was undermined.

Instead, he proposed that the interaction between *A* and *Non-A* produced a synthesis or new truth, thus becoming the *A* of the next

Diagram 2

TRADITIONAL LOGIC

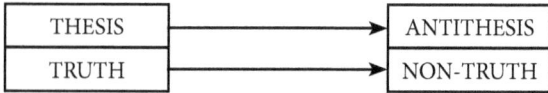

| THESIS | ⟶ | ANTITHESIS |
| TRUTH | ⟶ | NON-TRUTH |

THE DIALECTIC PROPOSED BY HEGEL AND PROMOTED BY MARX

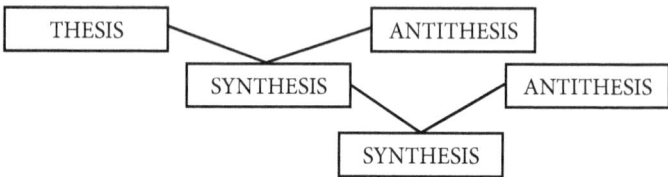

THESIS	ANTITHESIS
SYNTHESIS	ANTITHESIS
SYNTHESIS	

ongoing synthesizing process. In other words, the integration of *A* and *Non-A* would create a new interfused *A* that would consequently be integrated into the next level of this evolutionary process. In that way, truth is ever-evolving and never absolute; truth, then, is relative.

> Before his time truth was conceived on the basis of antithesis, not for any adequate reason but because man romantically acted upon it. Truth, in the sense of antithesis, is related to the idea of cause and effect. Cause and effect produces a chain reaction which goes straight on in a horizontal line. With the coming of Hegel all this changed. . . . Instead of thinking in terms of cause and effect, what we really have is a thesis, and opposite is an antithesis, but the answer to their relationship is not in the horizontal movement of cause and effect, but the answer is always synthesis.[3]

Karl Marx then extended this relativistic concept into the political realm, and from the dialectic process, the political theory of communism was developed.

Georg Wilhelm Friedrich Hegel   Karl Marx

Admittedly, some things are relative, and one cannot say that the process of the dialectic is totally untrue. The dialectic process may be useful in the acquisition of certain useful insights, but when it is applied to the determination of all truth and relegated as the only absolutely truthful process from which one can derive truth, it is both self-contradictory and disastrous in its consequence.

While it is true that not all things go in a horizontal cause-and-effect direction, biological and psychological systems sometimes exhibit the circular feedback effect necessary to achieve homeostasis. However, these exceptions hardly constitute proof that all processes are determined through synthesis. And it is a far cry from declaring absolutely that all things are relative and that truth cannot be absolute.

The irony of the inconsistency of this absolute stance somehow escapes most people. How can they state with absolute certainty that truth is not absolute and then remain consistent with their premise? One cannot state with absolute authority that truth is relative under a relativistic premise without being self-contradicting. It is a blatant oxymoron.

Hegel accepted the evolutionary ideology as the foundation for the establishment of a sociopolitical theory that at its very root shares the fundamental occult theological doctrine that we can become gods. That doctrine is not, as they claim, a progressive worldview that promises a future utopia; it is a regressivist occult doctrine—"you will be like God" (Gen. 3:5)—that was used by the serpent in the garden to deceive humankind. It is the universal great lie promoted by all occult groups.

> Hegel believed that the history of humankind was the story of man becoming more and more rational and "achieving consciousness." To "perfect" humanity, all that was needed was a government that tamed the impulses of human nature for the greater good. This was Hegel's revolutionary idea of progress. . . . Hegel concluded that the world now stood at one of the most advanced stages of human history and that experts and knowledgeable persons should rule with the most perfect government and unlimited authority over the individual. Through the state and its rulers, in Hegel's "philosophy of history," man essentially became God on earth. This was the foundational principle of what eventually became known as progressivism.[4]

This concept is at its very heart the antithesis of the understanding of the Founding Fathers of the great American experiment in self-government who saw the fallenness of humans and their propensity to hunger for power as a thing to be protected against through the checks and balances they instituted in our government. It is also the very reason that progressivism seeks to centralize power and push for its unlimited authority to control the individual in every aspect of his or her life. The Hegelian relativistic ideology invariably leads to an authoritarian regime and the loss of a citizen's personal liberties.

The Consequences of Entering the Age of Vericide

The death of absolutes in our modern society has sadly stretched out its tentacles to affect all the major disciplines—from philosophy to music, to the arts, to science, to our educational system, to politics, and even to theology. Everywhere in our society today, we encounter the ramifications of the death of absolutes, from the filling station attendant to the corporate executive.

People in all walks of life are expounding statements and adopting a lifestyle directly or indirectly as a result of this view without realizing its inevitable negative and even lethal implications. We have entered blindly into the Age of Vericide, the age of the death of true truth. The Hegelian view of the evolution of truth has so thoroughly permeated society that most people live and act in this system totally oblivious to its degrading effects on their individual lives and on society as well.

Humankind must understand that it is this evolutionary view of truth, which becomes the foundation of atheistic political systems that enables them to rationalize the inhumane treatment of humans. The death of absolutes is the death of morals and justice and the inevitable springboard to the abuse of human rights. Where there is no absolute basis for morals, there can be no social justice. In a naturalistic system, the weak become valuable only in regard to the measure for which they have any utilitarian value to the elite in power.

And why not since, after all, the evolutionary paradigm is the survival of the fittest? Hence, might determines right, and it stands to reason that the logical conclusion of this premise is that the weak are to be either eliminated from the gene pool if they have no utilitarian value to the power elite or exploited by the powerful.

Karl Marx embraced this Hegelian-evolutionary idea that humans could "progress" toward a more perfect form of government that would be scientifically administered from a centralized and absolutely authoritative bureaucracy that controlled every aspect of society. The brutality that resulted from this inhumane ideology was made evident for all with eyes to see when the Bolshevik Revolution of October 1917 established the Soviet Union. The oppression and violence it unleashed on its citizens may never be fully known this side of heaven.

The inevitable result of a culture that accepts the worldview that the determination of truth and morals can only be derived from pure reason on a strictly human or horizontal plane is the eventual and inevitable establishment of a society without equality of opportunity that panders to the needs of the elite at the cost of the weak. This is what is at stake, not only in America but in every other nation of the world.

Faith vs. Reason

Progressivists love to paint themselves as rational and tolerant multiculturalists. That is nothing more than another clever smokescreen. Those who have accepted the Hegelian relativistic worldview as their foundational ideology are vehemently intolerant of the Judeo-Christian worldview because they understand it to be the exact antithesis of their atheistic ideology and therefore the greatest obstacle they must overcome in order to gain political power. Their attempts to discredit the Judeo-Christian worldview are usually cloaked in terms that depict science and reason as the antithesis of faith. Faith is now equated with superstition and consequently categorized as the antithesis of reason. But is reason really the antithesis of faith? Is faith opposed to reason?

On the contrary. True faith cannot be derived without true reason. Is there anything wrong with reason? No, a thousand times no. True truth is not in contradiction to true reason. Rather, true reason leads us to true truth.

Postmodernists have divided the field of knowledge into two nonunifiable parts. On one side, they have the physical realm guided by reason and science. On the other side, they have the spiritual realm guided by irrational blind faith. A great gulf separates the two. In that way, faith is relegated to the mystical and irrational sphere, which has no communion with the rational process of induction. The rational side is relegated to the realm of science and reason, while faith is considered nothing more than subjective superstition.

But this division is also a smokescreen for their relativistic theology. That realm of science in their divided field of knowledge is exclusively the domain of Darwinism and naturalism, which are, in fact, metaphysical interpretations of the scientific data. Darwinism and naturalism are not empirical facts but rather a contrived theory that interprets those facts through an atheistic metaphysical grid.

This Neoplatonist dualistic worldview, which unnaturally divides reality into two halves, lends quite nicely to the relativistic worldview. But it is nothing more than an irrational leap of faith that modern man takes in order to remain within the naturalistic framework while holding some mystical faith to fill the void left in his or her soul by the sterility of the naturalistic paradigm.

The naturalistic paradigm sees reality as an impersonal series of cause-and-effect reactions guided by random chance. The natural outworking of this worldview makes the universe an impersonal series of chemical reactions, and humans are nothing more than organic machines. The divided field of knowledge makes our physical reality a cold and heartless random accident of nature. Humans can think that way in their brains, but they instinctively crave some transcendental significance. This unnatural division of reality does not pass the litmus test of reality because humans are hardwired to have a transcendent need. They were created in God's image.

The unity of true truth is a basic premise of the Judeo-Christian worldview. The physical and the spiritual are both parts of the same reality. The three visible spatial dimensions described by the superstring theory are just as much a part of the universe as the other seven invisible spatial dimensions. All 10 spatial dimensions are unified by the dimension of time. This is mathematically deduced and not a matter of blind faith. The rational science of mathematics unifies all aspects of reality—the visible and the invisible—in one equation.

But human reason alone is insufficient to develop the substratum from which absolute truth can be derived. If there is not a higher being than humans, the ultimate paragon, who can become the hitching post or standard for morality and truth, then there is no such thing as morals or truth. Therefore, from a purely horizontal perspective, humans cannot arrive at true truth. What is, just is and cannot be called either good or evil. Therefore, from a purely horizontal perspective, humans cannot arrive at any real standards for morals, justice, or the transcendental value of human life, which makes the concept of individual freedoms an illusion.

The real question behind this dilemma of whether or not truth is relative is whether or not there is a God. If there is no God, then humans are the final authorities who determine truth and morality through their own finite reasoning faculties on a horizontal plane. Each one chooses individually what is right and wrong, truth or lie, good or evil, just or unjust. In such a system, all choices are equally valid. Or, to be more accurate, they are equally invalid. There can be no measure of validity where all is essentially sameness.

It is because our culture has moved away from the idea that God exists that postmodernists have become intellectually arrogant, presuming that their finite brains can be the final authority in the determination of truth. It has led them to ridicule those who espouse absolutes as anachronistic and antiquated moralists. The word *moralist* is now a derogatory term that implies a narrow, closed-minded person who is unthinking and unenlightened.

In the place of God's moral standards, they have attempted to fill the void with the pragmatic concept of situational ethics. But the very concept of situational ethics is an oxymoron, for ethics cannot exist in a vacuum. Where there is no God, the word *ethics* is meaningless. All actions are simply neutral where there is no absolute standard to measure against. All that exists is power. All actions are permissible and morally neutral. Pragmatism and power become the true vehicles of change in such a culture. For that reason, progressivists, who see themselves as more enlightened human beings, are adamant about forcing their elitist ideology on those they consider unenlightened.

Their attempt to fill the void with their own perceptions of correct ethics is an unwitting admission that ethics are required in our society. This instinctive movement is the intrinsic evidence found in all human beings that belies the image of our creator in us. We all possess an intrinsic need for justice. But without a personal God at home in the universe, the very idea of ethics and justice is nonsensical and merely delusional.

We can see this tendency in the tactics of progressivist radicals who have been working to transform our nation since the 1930s. Saul Alinsky is a prime example. Alinsky's work became a training factory for so-called community organizers who were thoroughly indoctrinated in the Hegelian relativistic ideology. The impact that these community organizers have made on our political system cannot be underestimated. In fact, our nation has not only voted for one of his disciples to be our president and for Hillary Clinton to be the Democratic Party nominee in 2016, but it has also embraced Bernie Sanders, an avowed socialist. At the time of the finalization of this book in 2019, we have Democratic Party aspirants to the White House who are so far to the left that they make Bernie Sanders look conservative. The work of such as Saul Alinsky is now flowering in our nation like never before. Explaining this ideology of community organizers, Alinsky wrote, "To begin with, he does not have a fixed truth—truth to him is relative and changing; *everything* to him is relative and changing. He is a political relativist."[5]

This is not merely a scientific matter or even just a philosophical or moral matter. It is the underlying ideology that undergirds and fuels the Hegelian collectivist ideology that has brought so much death and suffering to countless nations in our world, including the Soviet Union, China, North Korea, Cuba, and Venezuela, to name a few.

But do not imagine for one moment that this is not an American concern or that our nation is not in peril. President Ronald Reagan reminded us that tyranny is always only one generation away. If you are frustrated at the great polarization that is dividing our nation through the clever ruse of identity politics and want to know its root cause, all you have to do is read the words of Alinsky, the master community organizer. But you will notice that even though he clearly refuted the idea that absolute truth exists and staunchly declared relativism, he cannot help but make religious-sounding motives that imply that radicals are on the side of goodness.

> The organizer must become schizoid, politically, in order to not slip into becoming a true believer. Before men can act, an issue must be polarized. Men will act when they are convinced that their cause is 100 percent on the side of the angels and that the opposition are 100 percent on the side of the devil. He knows that there can be no action until issues are polarized to this degree.[6]

But goodness to radicals is the destruction of the Judeo-Christian worldview and the establishment of a collectivist new world order. To do this, they must first undermine the constitutional principles that were responsible for making our great nation the freest nation in human history. This they accomplish through "identity politics" by creating divisions and fomenting hatred between blacks and whites, between women and men, between the poor and the wealthy, between every subsector of our human family and community. If you do not believe me, read Alinsky's own words.

> *The first step in community organization is community dis-organiza*tion. The disruption of the present organization is the first step in community organization. . . . The organizer dedicated to changing the life of a particular community must first rub raw the resentments of the people of the community; fan the latent hostilities of many of the people to the point of overt expression. He must search out controversy and issues, rather than avoid them, for unless there is controversy people are not concerned enough to act. . . . Enter the labor organizer or agitator. He begins his "trouble making" by stirring up these angers, frustrations,

and resentments, and highlighting specific issues or grievances that heighten controversy . . . and so the labor organizer simultaneously breeds conflict and builds a power
structure.[7]

We have seen first-hand in 2016 the fruit of this polarization ideology in the streets of our nation as a wave of cold-blooded assas-
sinations of police officers claimed the lives of countless husbands,
fathers, wives, and mothers. This is not some empty intellectual
sophistry. It is at the very root an ideology that impacts every aspect
of human existence, and we as Jews and Christians must know how
to intelligently oppose it.

This ideology in essence denies the existence of a good and
righteous God and makes humans gods, rabid narcissists whose
egos cannot be ignored in their pursuit of power. If you think I am
sensationalizing this, read what Alinsky wrote:

The ego of the organizer is stronger and more monumental
than the ego of the leader. The leader is driven by the desire
for power, while the organizer is driven by the desire to
create. The organizer is in a true sense reaching for the
highest level for which man can reach—to create, to be a
"great creator," to play God.[8]

The existentialist or relativist simply makes a choice and upon
making that choice declares reality, giving the individual the
complete freedom to choose as he or she wills, whatever pragmatic
action that leads toward accomplishing the goal of attaining power.
But more importantly, it allows people the freedom to be consistent
with whatever mode of morality they choose as their personal reality.
They become their own gods. Again, we are back to the primary
occult theological doctrine that has, since the Garden of Eden long
ago, taught humans that they can become gods. It is the great lie that
brought death and suffering to humankind.

And so we see that there is a natural progression that begins first by establishing a relativistic ideology and then fomenting a polarization of our society in order to fuel the violence that can be used as a power base for the revolutions generated by socialist-progressive-communist ideologues who have bathed our planet in blood in a sordid greed for power. If humans are gods, then all actions necessary and pragmatic to attain personal power are permissible and can be labeled as good for the cause.

True Truth Descends from God

If God exists and only if God exists, then justice exists, and God is the final authority in the determination of truth, morals, and justice. Then, God's propositional truth revealed to us from a vertical plane is absolute and universal. True truth descends from God, the author of truth. Quite simply, without God there can be no true truth, and all things are relative.

Directly related to this is the logical conclusion that if there is true truth, then a human cannot be its creator. Truth (absolute truth), cannot then be created or, to use the Hegelian term, synthesized by humans. It is, therefore, left to humans to either discover and accept it through true reason or to deny and distort it through false reasoning. True truth stands independent of the human mind—outside of humans—whether it is believed or not.

If, on the other hand, all truth is relative, then one escapes the conviction of an absolute moral code. There can be no paramount standard of ethics or justice in a naturalistic system. The concept of morality would be meaningless since there would be no absolute standard from which we could ascertain what is right or wrong, just or unjust. All individuals would be left to choose what is right or wrong for themselves, and they would not have the right to impose their own morals on anyone else since no one would have any more inherent authority than anyone else.

Each individual's reasoning would be as valid as the next person's. And no matter how antipathetic the conclusion of one individual is to another person, no matter how sordid or distasteful, it cannot be disavowed or regarded as illegitimate. But what are the long-term logical ramifications of accepting a worldview of relativistic truth and morality?

The brute fact of the matter is that if there is no higher being who has proposed a higher law from a vertical plane, then all human attempts to create a legal system that could provide justice are superfluous. The popular cliché "if it feels good, do it," which became famous in the counterculture of the late 1960s in America, would then be absolutely appropriate.

This is precisely one of the real reasons this view has become so accepted in most modern cultures. It categorically relieves people of any responsibility to an absolute moral code and allows each person to choose his or her own philosophy of convenience. It is not reason that stands behind this worldview; it is rebellion and emotion, the desire to do as one pleases without feeling guilty. It is not reason; it is rebellion garbed in the cloak of rationalization.

In the 1960s, our own American culture heralded a liberation from moral prohibitions that were blamed on the so-called religious establishment. The upshot of that has categorically all but destroyed the institution of marriage and the family unit.

How many broken hearts have suffered the pain of being abandoned for another? How many children have been raised without the benefit of the love of two parents? How many girls have become pregnant before wedlock? How many human beings have been killed through abortion in order to legitimize this hedonistic philosophy? Not to mention the consequent rise in venereal diseases such as AIDS that has claimed the lives of so many of our youth and will undoubtedly claim millions more.

We are witnessing the fruit of this lethal vine that has spread its tentacles into every discipline of our culture. Is it a coincidence? The resulting exponential rise in violent crimes and the increase in drug abuse are directly linked to the deterioration of the family unit. I need not say any more about this. Others have amply documented it.

We can witness this violent trend even in our public schools. The deterioration of the educational system is now made evident as teachers struggle to maintain order in a violent generation that vents its rage by killing not only teachers but also their students. School shootings are becoming the new norm.

All these negative repercussions and many more horrible things are the consequence of the teaching of moral relativity. All cultures that naïvely and unsuspectingly adopted this atheistic premise have historically and consistently fared poorly.

Unfortunately, human nature is such that most of us follow blindly after the pipers of our culture. These are the social trendsetters, the artists, the philosophers, and the musicians who express their philosophies through their own particular mediums. Most of humanity uncritically accepts whatever popular trend our respective cultures dictate in order to fit in and be accepted by their peers.

We must say with great conviction that it is not reason that has led us into the Age of Vericide. Sadly, the brute implications of relativism are almost invariably unexplored by the majority of the public who has ignorantly accepted this fundamental shift in their worldview. It has become accepted because we have desired to be free from any moral constraints.

Those who, for their own personal reasons, champion this view myopically think that they are liberating humankind from anachronistic morals that repress them. They confidently tell us that by accepting the naturalistic paradigm, we can escape the neurotic confines that they claim are the result of these harsh moral restraints.

They have negatively labeled these absolute moral standards as repressive moral inhibitions that rob us of our freedom and vitality. Belief in God and in an absolute standard of morality is now considered by the educational elite to be nothing more than the anachronistic and regressive superstitious behavior of the unenlightened.

The enlightened person, the "new man" of the coming New Age, is now liberated from these antiquated social restrictions that they believe to be the cause of neurosis in our culture. This was the clarion call of the existentialists who introduced our culture to this new paradigm in the last century. The call for absolute freedom demands the death of absolute morals and becomes the unseen proponent of abuse and violence in our culture.

But is the acceptance of this new paradigm really the liberator of neurosis in our culture? The acceptance of the philosophic premise that there is no absolute truth means that all people, no matter how depraved, neurotic, or asocial, may choose what is morally right or wrong individually with complete impunity.

To assert that each person is the ultimate measure of what is right or wrong is at best completely naïve and at worst the recipe for complete anarchism and eventual utter chaos in a society. It is not the clarion call to liberty and freedom. It is the foundation from which tyranny can in due time gain power and build its tower over humankind and rule with impunity.

It is in essence the denial of the very existence of morality and consequently of the assurance of any social justice. For without a standard that emanates vertically above humankind, there is no foundation for the development of any consistent measure of social justice. There can be no structure from which we can establish and ensure equality for all people, regardless of their color, nationality, gender, social status, or age, if there is not an ultimate paragon of truth, righteousness, and justice.

Relativism and Academia – The Problem of Indoctrination vs. Education

It is in the centers of academia that the future of our nation will be molded. Our centers of public education have instead become centers of indoctrination. An education requires the presentation of all sides of an argument. The student then has all the information necessary to make an informed decision. But our nation has been so mined by those who are pushing the relativist, progressivist agenda that schools have become monolithic and intolerant centers for indoctrination that no longer even feign to provide a venue for alternative viewpoints to be heard. They have ceased to be the catalyst of free thinkers.

It is almost impossible to become a tenured professor without towing the line of the accepted paradigm of materialism, and conservatives have also been denied the opportunity to exercise their God-given, constitutionally protected right of free speech on campus. Left-wing radical groups trained by community organizers actually riot when a conservative is given the opportunity to speak on campus. Our children are being brainwashed, and we have been intimidated from standing up and demanding our God-given rights.

Our educational institutions are in dire need of our intervention. Our federal dollars should be kept from institutions that refuse to educate and instead choose to indoctrinate. The vast majority of American academia has accepted a materialistic worldview as a de facto religion and refuses to allow any other competing worldview to be heard in educational institutions. In fact, that prohibition is extending even into the public square as the mainstream media colludes with the progressive agenda. They have relegated the Judeo-Christian worldview to the backwaters of our culture and have imprisoned us as anachronistic and unenlightened moralists.

The irony of it all is that their propaganda for relativism has been branded as freedom of choice. In reality, a student cannot make a choice if he or she is only taught one alternative. Moreover, going deeper into their underlying motivation for this slogan is their desire to be morally autonomous.

It is quite easy and popular to unthinkingly brandish the slogan "freedom of choice." But if truth is relative, then freedom of choice is absolute, and consequently, society would have no right to punish criminals because that is what they believe is right for them. Taken to its logical conclusion, it is evident that this freedom of choice must not be absolute, or the weak would lose their freedom by the selfish choices of the strong. This would therefore inevitably result in anarchy, which is eventually followed by tyranny. The absolute freedom of one individual cannot be attained without the loss of freedom for all others.

It does not take a rocket scientist to figure out that this would inevitably result in unfettered anarchy and absolute chaos. Who or what gives any person the authority to impose a set of standards on another human being intrinsically equal? For in a naturalistic system, each individual's standards are equally valid. In such a horizontal system, truth, morality, and justice are simply illusions. All that really matters is power.

All of us watched in utter dismay as the newsreels of the tragedy at Columbine High School sank in. We reacted with incredulity and utter helplessness as scenes of this gruesome event reached us. Two children who had accepted this Hegelian concept of reality went on a planned and well executed killing spree, killing other children, especially if they were Christians and stood for an ethical absolute that these individuals despised.

When a socialist-indoctrinated 66-year-old man deliberately gunned down Republican members of Congress practicing on a softball field in Alexandria, Virginia, he was being consistent with the relativistic ideology that believes the ends justify the means. And in fact, from a historical perspective, every book written by collectivist revolutionaries promotes violence as the means to conquest in exactly the same manner.

But why did the media act so surprised at the calamity at Columbine? Were these supposedly disturbed kids not consistent with the philosophy they were taught in the public school system? Why were they so surprised that the 66-year-old James Hodgkinson from Illinois was trying to herd all the Republicans to a corner of the softball field where he could kill them one at a time? If truth is relative and elitist, occult views are just as true as any other worldview, then it stands to reason that these people cannot be condemned for their actions. Nor can they be condemned for their bigotry.

Their truth is as valid as any other truth. And their moral choices accordingly are as valid as those of the president of the United States, Hitler, or Mother Teresa. The 19-year-old expelled student who went on a rampage and randomly killed 17 innocent human beings at Marjory Stoneman Douglas High School in Parkland, Florida, cannot be condemned for this brutality if truth is relative. Atheists have a problem. They cannot define evil. And yet in our hearts, each and every one of us knows that evil is real. Why, if we are nothing more than a serendipitous chemical accident of nature and the matrix of reality, is there a survival of the fittest?

The media elite have negatively ridiculed Christians who believe in moral absolutes, calling them moralists. But they cannot seem to see the glaring hypocrisy in their reaction to those who have lived out their terrorist worldview in such a violent form. On what basis do they condemn their actions? Materialism or relativism has no solid foundation from which individual freedoms can be protected. Their answer to this problem is collectivism, and all it leads to is the tyranny of the majority and the end of individual rights.

It is the intransigence of ignorance that deceives them into believing that their reasoning is unassailable. True reason is never in conflict with faith when there is true truth.

The Tyranny of the Majority – The Collectivist Ideology

Some have pathetically tried to escape the moral dilemma of describing what is evil in a naturalistic ideology by saying that what is right or wrong is relative to the majority opinion and not the individual. Society, in its view, makes a social contract for the good of the collective. The question then is this: In a relativistic system, by what universal or intrinsic authority does the majority opinion become an absolute?

In essence, they are saying that might determines right, not right determines might. They propose that the collective democratically determines a moral system in order to function as a society. And therefore, society forces its view on every member through a legal system. But if truth is relative, then the opinion that truth is dictated by the majority consensus applies only to those individuals who accept that view. It is not applicable to the individual who rejects that view. Hence, even in a relativistic system, a given set of moral dictates is always forced on some who do not accept that standard as valid. Then, where is this freedom of choice they speak of?

Under this presupposition, one would not have been able to condemn apartheid in South Africa or the abuse of civil rights in America. Under this system, those individuals who believed in the oppression and exploitation of minorities would have the right to do so since that is their truth. What right would we have under this relativistic system to impose any view on another human being?

Force, then, is the only solution available to their problem. But from a utilitarian point of view, it can be said that force may work to impose a certain standard, but does it legitimize it? Clearly, if majority consensus and power dictate right and wrong, then the atrocities of the Nazi Party cannot be called atrocities, for they were dictated by the power structure. Under this collective view of relativism, Hitler's attempt to exterminate an entire race of human beings could not be called wrong.

Genocide would be their moral prerogative in a relativistic system. The calloused slaughter of millions of men, women, and children who were systematically murdered—and many of them cruelly used as guinea pigs for scientific investigations—could not be condemned. Some were herded like cattle into the infamous gas chambers, and others were allowed to dwindle away to skeleton-like creatures through forced labor and starvation. These unfortunates who were unlucky enough to escape being chosen for direct extermination in the gas chambers when they arrived at the concentration camps were allowed to live so the Nazis could harness their labor for work projects in their war effort to rule the world.

The fact that they were, from the Nazis' evolutionary perspective, considered subhuman or of inferior genetic stock allowed the Nazi taskmasters the rationalization necessary to justify inflicting barbarous cruelties upon them. But again, this was not peculiar to the Nazi Party. The millions of humans who were killed in the Soviet purges dwarf the number killed by the Nazis. And the millions of humans killed by Mao's purges dwarf the Russian numbers. Yet each of these regimes rationalized these acts as necessary for the good of the collective.

Hence, under the Hegelian relativistic framework of philosophy, they were perfectly moral and right to do whatever they deemed proper for the survival of their regimes. How many millions of other human beings who escaped execution have been unjustly incarcerated and their lives completely ruined? Can we truly call this social justice?

***

The clarion call to freedom espoused by the existentialist is a diabolical ruse that can lead only to tyranny! The millions of ruined lives and murdered innocents left in the wake of political structures built on the naturalistic paradigm cry for justice from the blood-soaked ground of our wilting planet.

***

What we now state so quickly and superficially in one paragraph entails boundless individual cases of unimaginable misery and years of anguish created by the acceptance of this naturalistic premise by regimes bent on liberating the masses from the Judeo-Christian moral constraints.

Can mere words suffice to convey the depth of anguish and pain felt by just one man when his life's work is shattered in one single day? When all that his family before him struggled to obtain and passed on to their posterity is taken from him? When he is robbed from passing to posterity the fruit of his labor? When he is dragged from his home and family in manacles and thrust blindfolded in front of a firing squad? When he hears the countdown and the firing of guns, expecting at any moment for the bullets to tear through his flesh? When time stands still and he hears the sound of the gun? When, after a few numbing seconds, he realizes that the bullets whizzed by overhead and were only meant to terrorize his mind?

Can mere words convey the true depth of misery experienced by those who have been cruelly robbed of their freedom? Can we from the comfort of our hearths even begin to imagine the anguish suffered by such a man when he is beaten and tortured and left naked and cold in the middle of a dirty, damp, and miserably hot cell? When the companions in that cell are roaches, rats, and lice, and a suffocating and relentless tropical heat makes every breath labored?

Can we feel what he felt when he is tried in a kangaroo court and sentenced to years of imprisonment, simply because he does not agree with the party line of the elite in power? When he is ripped from his children at the prime of his life and denied the opportunity of shaping their lives and sharing in their accomplishments? When he is denied the warmth of his wife's love, the touch of her skin, the caress of her lips, the smell of her hair? When he is robbed of his health and vigor through the heartless abuse of the prison guards and the poor diet he is given because he is deemed an enemy of the state?

Such was the fate of a single man, a man named Francisco Jose Patiño (Paco), who happens to be my father. And is there a way to measure the pain and anguish that, like ripples in a pond, affect all those who love him? I know that pain from the inside out, but I am just one man. My father was one of the lucky ones. He lived to speak of his experiences in a communist prison. He lived to teach his children the value of standing up for what you believe, no matter the consequences. But what is the value of integrity in a naturalistic system?

How futile are words to convey the depth of anguish that just one single soul experiences when reason is profaned into rationalization and used as a weapon to mask the greed and selfishness of humans for power. Is it possible to explain to those who have not suffered the anguish of such injustices? Sadly, I think not. It seems that most people would rather not think about these things. It seems that most people would rather stick their heads in a hole in the ground, like the proverbial ostrich, and cross their fingers hoping that all will turn out right.

They do not want to be bothered by anything that requires any form of sacrifice or commitment. They are more concerned with material gain and pleasure. But they are blind, and soon their freedom will vanish, and they will wish they had taken action when there was still time to act. That time is growing very short.

We Are Our Most Dangerous Enemy

Our greatest enemy is not outside our borders; it is within our minds. Selfishness, greed, and apathy are the true nemeses of our nation. We turn our insensitive faces from the suffering of our brothers and sisters and choose instead to be entertained. How shallow our culture has become! How surprised we will be when that suffering comes knocking on our own door. This is the fruit of the Age of Vericide—the age of the death of truth, the age of relativism. And the fruit of it can be counted by those killed in our own public schools—the schools that have removed God from the classroom. Violence in our schools—the fruit of their relativistic ideology—is accelerating. I cry for the parents and siblings of those students who have been taught the Darwinist, materialistic ideology that human beings are nothing more than the product of chemical serendipity lacking any transcendental significance.

And yet this unjust suffering, endured with dignity by these parents, is but one grain of sand on an endless shore where the waves of injustice are crashing like giant tsunamis, crushing all within their path when the matrix of their ideology has become the survival of the fittest. The oppression from the political systems that have stemmed from this materialistic ideology is there for anyone to see in the history of each of these nations.

It is hard for Americans to imagine what it was like to live in a time and place where long sleepless vigils in the night were fueled by the ever-gnawing fear of that dreaded knock on the door from the secret police. But this is not as far away from us as you might be thinking. That sense of peace and safety that Americans have enjoyed for so many blessed years is now beginning to disappear since the rise of surveillance technology, the political weaponization of our intelligence community, and the radical terrorism coming from the Middle East.

The long-prophesied time of lawlessness is almost upon us. For many years, America felt safe, surrounded on both sides by vast oceans. That is no longer the case. Our threats now come from within and without. The destruction of the Twin Towers witnessed by American citizens on television has changed that illusory bubble, but that is only the beginning. One is coming who will take peace from the entire earth and embroil it in a time of lawlessness.

In spite of this, few in America can even imagine the terror that awaits them when our own government becomes an instrument of lawlessness and terror. This will be the inevitable result of the centralization of power engineered by the progressive ideology that has already created a massive federal bureaucracy that must be whittled down to the intended purpose of our Founding Fathers. We are already witnessing the collection of big data in our homes by wonderful little devices that have names and to which we can ask anything and get an answer. The only problem is that these devices are recording every word we say inside our homes and establishing a tremendous data base that could be used quite effectively by a tyrant.

Power must not be allowed to be centralized, or corruption will become the natural spawn of this grave mistake. That knock in the middle of the night does not yet register the emotions that one day will be more clearly perceived. It is that knock that has torn millions of families apart, never to be reunited again, all over our planet. It is that knock that took your father or brother or sister or mother to a field where they were callously shot with a bullet in the head or the back of the neck and then kicked into a trench. It is that dreaded knock where they took your father or brother or sister or mother to a Siberian gulag to be raped, tortured, and eventually worked to death.

Why were they taken? Not because they were criminals or murderers or disturbers of the peace. Not because they were violent people, but simply because they were deemed a political threat to the centralized state. They may have owned a little more land than others. They may have been educated and knowledgeable enough to see through the propaganda of the state. Or, like in Estonia and Georgia, they were just numbers in a quota sent by the central government that was trying to winnow down the native population so they could backfill the land with Russians in order to claim that land permanently.

This is not ancient history. This is what is happening in Crimea today, and I fear it will be happening soon in Ukraine. Almost half of Ukraine is now in their grasp. It is what is happening in Syria as ISIS attempts to kill every Christian and Kurd. It is happening in Africa as radical Muslim groups attack churches and schools and capture the young women to be used as sex slaves. Genocide is alive and thriving on planet Earth. Few have survived these ordeals to live and tell about them. My father did. Solzhenitsyn did. We should listen to their warnings.

Most of us instinctively cringe at such injustices. "How could anyone be so cruel?" we say. But there is no such thing as cruelty where there is no God and where the matrix of reality is the survival of the fittest. This is the only terminus for the Age of Vericide. This is the direction our nation will be following if we do not intelligently win our children to the truth and inoculate them from the deceptive propaganda of the collectivists disguised as progressivists and even radical Islamists bent on the colonization of the entire planet.

The Hegelian relativistic framework of morality has been applied to the political process by many nations, and the inevitable result has consistently been cataloged for those who care to see. In each and every case, the sanctity of human life has been ignored and trampled upon by those who rationalized these atrocities that were necessary in their struggle for power and control.

Genocide was and is considered a necessary action deemed absolutely legitimate for the pragmatic subjugation of any territory for takeover by occupiers and usurpers. Those who hold to the demonically inspired globalist aspiration, whether in the form of radical Islam or a collectivist ideology, have never and will never hesitate to use genocide as their most pragmatic, effective, and long-lasting policy to gain complete control of the lands they colonize.

But even those who survive such atrocities are changed from within when they become seduced by the collectivist ideology. When the laws of a nation are based on a relativistic system, greed becomes the norm. The government wants power, and the people want free things. Their greed shackles them to the government, which will be their master, and they will become its slaves. The people's greed blinds them to the fact that they are selling their birthrights for a pot of porridge. The government cannot give you anything it does not first steal from someone else. The problem is that eventually, the government will run out of people to steal from.

"That will never happen here," you say. Oh, how I wish that were true! We have already begun to see this overreach of power when the executive branch of our government used the Internal Revenue Service (IRS) to go after its political rivals. We see the abuse by progressive courts that have steadily whittled down our religious rights by ignoring the fundamental doctrine of the separation of powers that our Founding Fathers created to protect us. Both the executive and judicial branches are now in the progressive habit of legislating. They have robbed the people of the United States of that right. It is supposed to be "we the people" through our representatives who have the exclusive right to create new laws.

As a result, they have murdered the unborn *en masse*. I warn you, that disregard for the value of human life will not end in the womb as the Darwinist-materialist ideology begins to change our culture from the Judeo-Christian ideology upon which it was founded. Progressives already consider conservative Christians greater enemies of the country than radical Islam. They have tapped into our phones, computers, and every financial contract we make. They are beginning to collect samples of our DNA. Cameras are now placed all over our cities and roads. Vehicles and cell phones can be tracked via satellite.

The technological abilities of modern governments to rule with an iron fist as never before are in place, waiting for Satan's moment to strike. We have already seen the politicization of the FBI and the Department of Justice used by the executive branch against a political opponent. Those who in their progressive-socialist mentality think all actions can be pragmatically rationalized to impose upon us "unenlightened" members of society believe their worldview is legitimate.

Today in our nation, we have an openly declared socialist who ran for president of the United States of America and may have won the Democratic Party candidacy had Hillary Clinton not put a fix on the Democratic Party. Not that she would have been that much better for our nation—she has the same ideology as Bernie Sanders, just on a slower path.

Sadly, most of the youth who have been "enlightened" by our public education system are enthusiastically supporting Sanders. They have been thoroughly indoctrinated in the relativist-collectivist ideology, and the day will come when they will be the ruling majority—that is, if it has not already come. Listen to the words of Solzhenitsyn as he describes the consequence of the Hegelian system in the Soviet Union. Like Elijah crying in the wilderness so many years ago, Solzhenitsyn tried to warn us of the danger of the collectivist ideology.

This was the system that, in the twentieth century, was the first to introduce the use of hostages—that is to say, to seize not the person whom they were seeking, but rather a member of his family or simply someone at random, and to shoot him.

Such a system of hostages and the persecution of families exists to this day. It is still the most powerful weapon of persecution, because the bravest person, who is not afraid for himself, can flinch at a threat to his family.

This was a system which was the first—long before Hitler—to employ false announcements of registration, that is to say: "Such and such persons must appear to register." People would comply and then they were taken away to be killed. For technical reasons we didn't have gas chambers in those days. We used barges. A hundred or a thousand persons were put into a barge and then it was sunk.

This was a system which deceived the workers in all of its decrees—the decree on land, the decree on peace, the decree on factories, the decree on freedom of the press.

This was a system which exterminated all other parties. And let me make it clear to you that it not only disbanded each party, but destroyed its members. All members of every non-Communist party were exterminated.

This was a system which carried out genocide of the peasantry. Fifteen million peasants were shipped off to their deaths.

This was a system which introduced serfdom, the so-called passport system.

> This was a system which, in time of peace, artificially created a famine, causing six million persons to die in the Ukraine between 1932 and 1933. They died on the very threshold of Europe. And Europe didn't even notice it. The world didn't even notice it. Six million persons! (emphasis added).[9]

Six million human beings were purposefully starved to death, and even America hardly noticed it. Can we afford to hide in the comfort of our cozy homes and not notice the wake of death left behind each of these political systems that has adopted the relativistic framework as the matrix of reality? Do we dare to imagine that such a system will not produce in our country the very same results it has consistently produced in every other country in which it has become entrenched? I am not a fearmonger. I am not a warmonger. I am a realist, and if America does not remove her pink progressive glasses soon, there will be no return. Relativism is the direct road to tyranny used by monsters of every political spectrum.

Hitler's occult-atheistic ideology accepted the same relativistic and pragmatic view of human life and effectively promoted it through slick propaganda by the National Socialist ideology. Human life was regarded as a mere utilitarian instrument for the attainment of power for the elite—the Aryan race. Today, under the presupposition of relativism, Hitler can only be considered a monster not because of what he did, for he was being consistent with his philosophy, but because the Allies won the war.

Relativism simply cannot create a structural foundation for the sacred individual rights of human beings. It is instead the direct path to tyranny and abuse.

Individual Human Rights

Our American fundamental human rights do not depend on the outcome of an election, on the authority of a federal judge, or on

the permission of a humanly contrived system of government. It is not dependent on the proclamation of any politician or ruler. It is not even dependent on the will of the majority. Our beloved and sacred Bill of Rights created at the very inception of our U.S. Constitution drew its legitimacy from the foundation of the higher document, the Declaration of Independence. And this is the primordial foundation upon which all our human rights rest because every human being has been created in God's image. These rights are intrinsic to all humanity. They stem directly from the creator, who is the supreme judge of every nation and people. This is the only real basis for human rights in every nation of the planet.

This is the bold and rational foundation upon which our Founding Fathers anchored our human rights with the express intention that it should become a perpetual, eternal, and unchangeable foundation. And because God's authority is above the authority of any human construct or government, they declared openly that these rights cannot be legitimately abrogated by any form of government—past, present, or future. In fact, from the onset, they stated that if any government dared to usurp those rights, it was the duty of every citizen to revolt against that government. That is the backbone of the American experiment.

That backbone has now been riddled with ossification by the cancerous acceptance of relativism as the most popular American worldview, especially among our youth. If humanity is a product of an impersonal evolution—the mere product of matter and energy, space and time in an impersonal chance cosmic accident—then humans are nothing more than impersonal composites of chemicals without any transcendental significance whatsoever. They are simply incredibly sophisticated amalgamations of chemicals that, by purely blind or random chance, have luckily evolved into an autonomous, highly skilled, self-programming, biological computer.

Human life, then, has no more inherent value or meaning than any of the other "accidents" that coincidentally evolved alongside human beings. Human life is therefore of no more intrinsic worth than any other animal or, for that matter, an inanimate rock that has evolved alongside it. Humans have the inconsequential distinction of being no more than organic machines!

Relativism provides no firm foundation for securing individual human rights. Quite the contrary, it provides the basis for the abuse of the weak by the strong. Sadly, this view is being promoted not only by our educational system but also by the mass media and the film industry that is filled with perverts who abuse young starlets, both male and female, with lying promises.

The gargantuan hypocrisy assumed by these movie stars who have no moral compass and yet seek to presume they can use their fame and stardom to guide us into what is correct leaves me speechless. How do they assume they have such a pulpit? What have they done or sacrificed to help the world and the poor? They are simply famous for being famous. Their narcissism is exceeded only by their ignorance of the true truth.

In Steven Spielberg's movie *A.I. Artificial Intelligence*, this naturalistic idea is masterfully addressed. The audience is brought to the conclusion that machines or robots are the final step in the evolutionary ladder of life. The movie spins a tale of an adopted child robot or human android named Dave who was created to take the place of a child who, unfortunately, suffered from a terminal disease. The ill child (Martin) was therefore preserved in a suspended state of animation through cryogenics.

Initially, the mother rejected the android Dave. But we are led to believe that the technological advancements had reached such a level that this new type of robot, called a Mecha (mechanical) android, had been programmed to be capable of loving. Naturally, his love for the mother eventually won her over.

The mother, suffering terribly from her loss, eventually filled her vacuum with love for Dave. But in time, the human child was thawed out after medical advancements were able to restore him to health.

Subsequently, the Mecha child Dave had a sibling conflict with the human (Orga) child Martin, and Dave was consequently slated for destruction. But the mother could not bring herself to destroy Dave, having become emotionally attached to the Mecha child. Instead, she released him into the woods to fend for himself.

In the story, we are led to believe that Dave was successfully programmed to love his mother. And, as in the story of Pinocchio, the plot revolves around the robot's quest to find Blue Fairy, who will turn him into a real boy, an Orga. The robot child reasoned that if he could find Blue Fairy to turn him into a real boy, the mother would equally love him.

During his search for Blue Fairy, Dave is rounded up, along with other outdated and discarded robots, to be violently destroyed in an arena for the entertainment of humans. The scene is reminiscent of the Roman arenas where Christians were fed to wild animals for the entertainment of the drunken crowds. The heartless destruction of these robots in gruesome theatrical conventions automatically elicits the viewers' disgust for this type of insensitivity in humans.

In the end, the human race has disappeared from the face of the planet, presumably through their violence. "Living robots" are all that is left to inhabit the world. Thousands of years later, "highly evolved living machines" find the little robot (Dave) and take pity on him. They grant him his wish by recreating the mother from the DNA of a lock of her hair. For one day, he gets to be with the mother again and finally feels the love he had so longed for in his protracted and arduous quest.

This well made film incites within us emotions of pity and compassion toward the Mecha child who was treated so unjustly and rejected by his human parents and the inhumane Orga society around it. The audience is cleverly duped into thinking that these robots are, in fact, superior in morality to their human creators, or Orgas.

Most who watch the film uncritically accept that machines could really display the deep emotions of love and pity that are exclusively the characteristic nature of humanity. Sadly, the audience is cleverly deceived into subconsciously adopting the presupposition of the naturalistic evolutionary paradigm without understanding the horrible implications that stem from this worldview.

Our natural empathy for the abused and downtrodden is cleverly and calculatingly manipulated to accept the moral superiority of machines over humans. But if viewers could step back and analyze this message objectively, they would realize that the very reason audiences have these feelings of empathy for the abused comes from the foundation of the Judeo-Christian worldview.

Of what value is empathy and compassion in a naturalistic system? Perhaps it is fair to say that this foundational Judeo-Christian worldview lies dormant within our now secular culture. But without it, the film could not elicit empathy from the audience. What could be morally wrong with discarding useless machines and humans in a morally neutral, naturalistic system?

Moreover, if there is no God, then there is no real difference in the intrinsic value of a computerized machine (Mecha) and a real human being (Orga). As a matter of fact, there is no substantial difference in value between the living and the nonliving, between life and no life, between a human being and a rock.

This is the precise mindset that is necessary in order for our society to accept the merger of machine and humans in the development of transhuman cyborgs that will become a reality perhaps within a few decades. I urge you to have great caution here, my friend, for this is not as farfetched as some may think.

If there is no personal God who created humans, endowing them with that qualifying, all-significant factor called *personhood* that consequently crowns them with intrinsic transcendental meaning, then humans are simply reduced to the worth of everything else that has supposedly evolved alongside them. But the sterility of a machine without a soul is antipathetic to our subconscious. And so, for this reason, Spielberg instinctively had to artificially attempt to give the machine the personhood observed in humans.

And so we imagine that we can give a machine the ability to love and have pity, while in reality, by the strict definition of the naturalistic presupposition, none exists. Love in a naturalistic system is merely the sum total of electrochemical reactions. And if the matrix of reality is the survival of the fittest, then it is the duty of the powerful to dominate and enslave the masses.

In *A.I.*, Spielberg crosses over the line of despair spoken of by Schaeffer into the irrational in order to assuage his transcendent need for significance. He cannot live in a mechanical world without love and must therefore superficially endow the machine with the humanity he desires. He must dress the cold metal robot with a ghost in order to make him more palatable to his subconscious transcendental need.

However, this is but a blind leap of faith that cannot be scientifically substantiated, and more importantly, it cannot be supported by his underlying worldview. It is the inevitable irrational leap of faith necessary to assuage the intrinsic transcendental need of humans for significance. For this is the way humans are, having been created by a personal God in the semblance of His image. Spielberg leaps into the irrational in order to assuage his intrinsic transcendental need that is in him because he was created in the image of God, no matter what his brain believes.

Even more incredulous is the point of the story, which Spielberg masterfully spins, that implicitly declares that in the end, these cyborgs are morally superior to humans. The irony of this irrational conjecture is that under his philosophic presupposition, one cannot derive any moral values at all.

If evolution is the matrix of reality and there is no absolute paragon of morality (i.e., God), then there is no standard from which we can arrive at any morality or sense of justice. The injustices of humans that he depicts in his film are recognized as injustices only because the audience is interpreting them through the Judeo-Christian grid of morality.

From the viewpoint of a relativistic framework, the utilitarian grid, which is the very reason the outdated robots are carelessly destroyed, is in fact, correct and appropriate. The feelings of disgust Spielberg precipitates in his audience as they watch the outdated robots being violently destroyed in an arena can only be elicited from those who think from a Judeo-Christian perspective, even if it is a distant memory.

In reality, the most someone can do is program the machine with certain preconditioned responses to outside stimuli that mimic personality. But personhood or even self-awareness cannot be imputed into the machine. But one could cheat by merging the Mecha with the Orga.

In the very near future, humans will be able to merge life with machines. But unless the hybrid cyborgs had spirits to begin with, they will not possess personhood, the intrinsic property of the human spirit endowed by the creator and inherent in every human being.

A person's soul, on the other hand, is the reflection of the interplay between the spirit and the body. It is personality (psyche). Our personality is subject to change, but our personhood is immutable and eternal. We can give a machine personality, but we cannot give the machine personhood. A cartoon has personality, but it does not have personhood.

We will address this topic again later. But for now, suffice it to say that the merger of machine and humans is just around the corner. Some in the transhumanist movement claim it will happen within the next 20 years or so. You may be surprised to know that a robot named Sophia has become the first robot ever granted official citizenship in any nation. The nation is Saudi Arabia.

Natural Morality

Some naïvely attempt to build an argument for morality based on observing Mother Nature. They propose that humans should duplicate the homeostasis obtained by nature and endeavor to participate in this harmonious ballad of life. But their inability to remain consistent with reality in a naturalistic system is unmasked by the very fact that they label the natural forces of a natural universe, which they claim has come into existence without divine intervention, as Mother Nature.

The rock we live on, called planet Earth, is referred to as Mother Earth. Their vain attempt to interject personality where none should exist in a naturalistic system belies their inherent need to personify what, by definition, they claim to be impersonal.

If there is no God and humans are alone in the universe, then we are simply the sum total of impersonal chemical and electrical impulses. Consequently, personality or, more accurately, personhood is but an illusion. But humans cannot live in the naked cruelty of an impersonal universe because they were created in the image of God. And therefore, for this very reason and only for this reason, they have the universal intrinsic need for transcendental meaning and significance that betrays even the most hardened atheist. They cannot escape their humanity because they were hard-wired by God. No matter how hard their brains try to deny it, their spirits betray them.

So those who attempt to avoid God still give nature, perhaps subconsciously, a personal persona. Nature, then, becomes Mother Nature. And Spielberg must inject the robot with personhood and the ability to love in order to escape the sterility of his underlying mechanistic worldview.

Here, Spielberg crosses the line of despair spoken of by Schaeffer and goes to the irrational in order to assuage his God-given need for transcendence. He is consciously aware that by strict definition under his naturalistic presupposition, there is absolutely no real significance to life or love or compassion. And yet he is subconsciously forced to make the machine more palatable to his intrinsic subconscious need.

But can Mother Nature give us a foundation from which we can develop a moral standard through reason apart from God? While it is true that the ecosystems in nature seem to display an uncanny

ability to remain in balance, just how does Mother Nature teach us about living in harmony? The harmony created by Mother Nature is not predicated upon peaceful coexistence. The present harmony displayed by Mother Nature is predicated upon predation!

And how does Mother Nature provide for us a system of protecting individuals against the powerful? Ask the zebra caught in the jaws of the lion about living in harmony. Or ask the wildebeest caught in the jaws of the crocodile twirling in the depths of the murky waters of the river while being dragged to the bottom, drowned, and torn apart. Ask the antelope being eaten alive by a pack of wild dogs as it watches its own innards being eaten while still conscious. Ask the insect caught in the spider's web, frantically attempting to escape while trapped alive and waiting for the spider to deliver its deadly bite that will dissolve his innards into a soup.

Pray tell, just how does nature teach us anything about harmony and equality and justice for all and individual rights? And how does the value of an insect differ from the value of a human being if there is no God? Is intelligence the measure of the worth of a being? Should we give a national IQ test and cull the bottom half of citizens to advance our evolution? Are we even able to say that nature makes us more valuable than rocks? Reason alone is incapable of providing us with the substratum for justice, equality, and morality, for in a godless reality, only power rules.

And what do we do when unwanted life forms such as insects invade our domain? Is it immoral to kill a roach? The realization of this implication is what allowed Hitler to say, "I cannot see why man should not be just as cruel as nature."[10]

Nature only teaches us that violence and force are the paramount prerequisites to survival and prosperity in the animal kingdom. We do not encounter the notion of justice for the weak or the rights of a minority in nature. There is only raging brute force, and the strong prey on the weak. The harmony of nature that atheists speak of is predicated on the abuse of the powerful on the weak to supposedly maintain a more robust genetic form by killing the weak. Are we to duplicate this in our culture? That is what Hitler proposed. Why would he be wrong?

Justice is exclusively a human concept that exclusively emanates from the following two Judeo-Christian doctrines:

1. The acknowledgment that all human beings are, in fact, brothers and sisters, children of the same common parents: Adam and Eve.

2. The direct consequence of the Judeo-Christian worldview that all humanity equally shares the God-given rights to life, liberty, justice, and equality of opportunity because they have been created in God's image.

Therefore, each human being is endowed with infinite worth. It is only from the Judeo-Christian worldview that the concept of *Lex Rex* (Law is King) could develop. The law can be king only if it is absolute, and it is absolute only if there is a personal God at home in the universe who has expressed this knowledge to humans. Relativism cannot become the foundation of *Lex Rex* since the law would then be subject to the interpretation of purposeful human machinations in synthesizing the product deemed most pragmatic and profitable for the powerful elite. Relativism leads back to *Rex Lex*.

Outside this foundation, there is no ground from which true and unbounded justice can develop and from which the sanctity of human life can be derived. The higher morals of selflessness and sacrifice for others are rarely found in the natural world, apart from humans.

By the way, it did not used to be that way when God first created animals. When sin entered the world, everything caught therein groaned under its weight. So what we almost exclusively witness in nature is fierce competition that often leads to predation created by the instinct for survival. When God first made lions, they ate grass like oxen. Now, lions eat oxen. But when the Messiah returns, He will change all of that again.

Also the cow and the bear will graze,
Their young will lie down together,
And the lion will eat straw like the ox.

—Isa. 11:7

Nature no longer reflects the original intent of God. We live in a fallen universe, and the second law of thermodynamics is the matrix of this fallen reality in each and every category of reality. The scriptures claim that even nature groans in this present travail, awaiting the time of its restoration due to the fall of humankind. And so do all of us, for that matter.

For we know that the whole creation groans and suffers the
pains of childbirth together until now. And not only this, but
also we ourselves, having the first fruits of the Spirit, even
we ourselves groan within ourselves, waiting eagerly for our
adoption as sons, the redemption of our body.

—Rom. 8:22–23

The Birthing of the Master Race

Hitler was quite clear about the implications of a naturalistic worldview, and his campaign to kill those he deemed undesirable was quite consistent with this philosophical foundation. As a matter of fact, higher human notions of pity for the disenfranchised and love for all humanity—regardless of age, sex, skin color, nationality, intelligence, or possessions—are, in a naturalistic system, a hindrance to the survival of the fittest and the evolution of humans.

These human sentiments cause us to protect, within the gene pool of society, genes that weaken the progress of this supposed evolution toward a more aggressive and successful species. It is therefore not surprising that the occult view of human evolution into godhood is quite compatible with this evolutionary concept, for they have a common author and a common goal.

In this, Nietzsche was quite clear, for he derogatorily viewed all fellow atheists who did not concede to the ideal of aggression and force, which he deemed to be the natural good, as cowardly, pale atheists. By this, he meant to say that they are cowards because they are not willing to face the true implications of atheism, which he called "facing the Minotaur."

He was absolutely right, for they are naïve beyond words and utterly at odds with reality when they interject anything other than naked aggression and brute force as the matrix of reality in a naturalistic world where there is no God. Any attempt to find middle ground is without a rational base.

In Hitler's view, life had no intrinsic or transcendental worth other than its utilitarian value. And since he considered Jews to be a detriment to his Third Reich, logically and as a direct consequence of the foundational presupposition of Hegelian dialectic reasoning, it must therefore be concluded that it was morally right and expedient for him to exterminate Jews, at least from his relativistic point of view.

His attempt to create a master race was no more than an attempt to accelerate the evolution of humans by annihilating from the genetic pool of our civilization those he deemed undesirable. He did with humans what we do with cattle and other domesticated animals when we try to create certain traits by carefully manipulating the breeding to select the genetic components that will produce our desired results, while eliminating those we do not want.

Hitler, through Himmler, chose the very best specimens of German youth to become members of an elite force he created to birth their superhuman race. Central to their ideology was the selection of specimens who were racially pure. In order to qualify as a member of the *Saal-Schutz* (or SS) guard, Germans were required to prove their pure Aryan heritage all the way back to 1750 AD.

These tall, blond men were instructed to produce as many off-spring as possible, with or without wives. German women were taught that their bodies belonged to the state and that it was their moral duty to produce offspring who would become elite members of the Third Reich.

Central to the Nazi occultic, naturalistic philosophy was the idea, promoted by Himmler and the *SS Ahnenerbe* (SS Ancestral Heritage Society), that members of this race that existed in the past were like gods. Most of them supposedly perished with the lost civilization of Atlantis. The Nazis were convinced that Aryans were descendants of this ancient race of god-men who were destroyed by a Gobal Flood when the Earth was struck by a moon.

According to Himmler, some escaped the deluge and were able to reestablish their race from the mountains of Tibet. For this reason, the Nazi high command sent scientific expeditions to the Tibetan mountains to find evidence that would substantiate their occultic claim. It was this foundational premise of Aryan superiority that birthed the idea of establishing a ruling race of superior human beings to rule the world.

There is, in fact, ample historical evidence of a race that was profoundly physically superior to common humans. These accounts are found scattered in the ancient mythologies of all cultures. The ancients refer to them in various ways that imply their superiority. Some called them titans, others giants or cyclops, but all of them considered this ancient race to be semi-divine.

No doubt they claimed to be gods in the same way that kings and other megalomaniacs, filled with the insolence and pride of demons, have done throughout human history. There is but one God in the universe. However, there seems to be a different twist to this ancient race of beings that separates them from the megalomaniacs of our age.

They may have been human beings of a different race that has now become extinct. There is good reason to believe that the extinct specimen of humans we presently call Neanderthal Man were, in fact, the Nephilim of Genesis 6. Their powerful, robust bodies and much greater cranial capacity were not the lumbering idiots that evolutionists would have us believe.

They had massive muscles and were intelligent, powerful beings who brought sorcery and the bloody religion of the serpent to our planet. The Greeks called them cyclops. The eye of the cyclops speaks of the third eye of the occult. It is a symbolic term to mean the ability to see the secret esoteric knowledge of the occult worldview.

So we see that the Nazi pagan religion was, in fact, an occult version of the biblical story of the Global Flood. Their attempt to reestablish the global control of Satan on the second earth was in continuity with the initial program of the Nephilim. And for this reason, they attempted to breed a new race of Nephilim. These I refer to as the Neonephilim, the race the Nazis sought to bring forth through their selective breeding processes.

A special program was instituted in 1935 to accomplish Himmler's goal of creating this special breed of superhumans called *Lebensborn*. It is estimated that 10,000–12,000 babies were conceived and cared for by the state through this program designed to create racially superior children. Most babies conceived by these SS guards for this purpose never knew their parents. Most were cared for by the state, although some were adopted by SS couples.

Unfortunately, these innocent children who, through no fault of their own, had been born into this system became social outcasts after the war ended. These sad children were referred to as "children of shame" and ruthlessly ostracized. Many of them, without justifiable cause, were sent to insane asylums where they suffered greatly. How sad we are when societies cannot understand that two wrongs do not make a right!

Perhaps one of the most famous of these Lebensborn children was Anni-Frid Lyngstad, known as Frida. She was one of the singers of Abba, the former pop cult band. She was born to a Nazi SS officer and a Norwegian mother. After the war, her grandmother took her to Sweden to escape mistreatment. Sadly, these children paid dearly for the crimes of the Nazis. But this was the least of the Nazis' barbaric accomplishments.

Their preoccupation with racial purity was such that even Germans who did not measure up to their standards of superiority were targeted for disposal. Parents of German children with any medical deficiencies were required to register them in government medical facilities at birth. They were monitored during their development, and if they did not outgrow their medical deficiencies, Nazi doctors secretly killed them with lethal injections. The parents were simply told their children died of natural causes.

Today, this process of special selection to help evolution toward the creation of god-men has been fine-tuned through our modern technological advancements. The process has now been perfected through the technology of recombinant gene splicing. We are now able to interject desired genes without waiting for multiple gestations to achieve the end result—and this without the added complications of undesirable genes being mixed in during the natural process.

The stage is once again being set to bring to our planet a special breed of humans who will usher in a global kingdom. Satan failed with Hitler, but do not for one moment imagine that his plan for earth has been completely derailed.

The violence that brought on the first earth certain destruction through the Great Flood will return to the second earth. The cruelty and violence of the occultic Nazi agenda did not die with Hitler. Those who pander the same pagan philosophical worldview are alive and well and entrenched in the higher echelons of power in our second earth. They patiently wait behind the scenes for the opportune moment to spring into action. From the Middle East, we face the colonialist-globalist ideology of Islam. From the West, we face the globalist ideology of those who hold to the Hegelian-naturalistic paradigm.

Those who espouse the Hegelian-naturalistic paradigm are bringing upon our civilization a future of violence and cruelty that humankind has never known. But can we condemn violence and cruelty if we accept the naturalistic paradigm as the matrix of our social order?

If there is no God, this cruelty cannot be condemned. As a matter of fact, the meaning of cruelty would not even exist, for the very definition of cruelty presupposes that unjust actions against a victim are morally wrong, and no such thing can be construed from an atheistic or relativistic worldview where, in the final analysis, all actions are neutrally amoral. Not immoral, but amoral. The naturalist is left without a platform from which evil can even be defined. How can we rationally condemn the evil that has been perpetrated against the innocents throughout human history when we cannot rationally define what is evil?

Therefore, reason is not in contradiction to true truth. Reason is not in contradiction to true faith borne out of true truth. And yet even the hearts of many who ignorantly claim to champion the naturalistic paradigm are instinctively repulsed by the horror of these wicked deeds. It is the palpable evidence for those with an objective mind that the naturalistic paradigm is incongruent with the way humans are.

It is the palpable evidence for those with an objective mind that all humans are, in fact, created in God's image. Try as they may, they cannot escape their humanity. This is the "mannishness of man," spoken of by Schaeffer, that is intrinsic to every human being.

But some of us are unwilling to face the truth. We are unwilling to face the Minotaur. Some of us hold on to the naturalistic paradigm, not because of reason but because of misguided emotional compulsions.

It is our selfish desire to be free of any moral restraints and to justify our greed for power or wealth or lust, which lies at the very root of our rejection of God. We have seen the outcome of such careless notions applied to political science and carried out with lethal consequences in the past.

We have seen the carnage that it produces. Will we allow it again in the future? Can we afford to be so naïve? Will history repeat itself? If we allow it, then it will undoubtedly again result in the same dehumanization of another sector of our human family. Already in our country, the unborn have been dehumanized.

Can we not understand that what awaits our culture if we do not oppose the entrenchment of this worldview are more trenches filled to the brim with victims left in the wake of the struggle for the consolidation of global power? Look carefully in the trench below, for you may one day be one of those cadavers if we fail to oppose the utter horror of the acceptance of this worldview in our present culture.

Be not deceived, my friend, or you may become the victim of your own ignorance. Most would view the above picture with disgust as they see emaciated bodies dumped into a mass grave. But they are unable to truly grasp the amount of pain and suffering individually experienced by each of these men, women, and children. Each one is a story of great tragedy and untold suffering that could fill a book.

Is the human mind capable of relating to such suffering when a person has not experienced suffering in any great sense? When I communicate with a profound sense of urgency the dire consequences that will befall us if we continue to ignore our responsibility to countermand the naturalistic paradigm, there is in me a sense of foreboding that is inescapable.

One of the many mass graves of the Nazis' Jewish concentration camps

There is a pain in my soul, a deep and sad realization that most of humanity will not listen. Sadder yet, even most Christians and Jews will not awaken from their magical enchantment. They are generally concerned only for the immediate gratification of their senses. Like Esau, they forfeit their heritage in the future for a pot of porridge today. They are addicted to the immediate gratification of the senses, and that is Satan's specialty.

I am conscious of the reality that most of humanity simply just will not care enough to lift a finger. But the power of a single finger may be more than we realize to accomplish good as it is also to accomplish great evil. To this one finger, I plead and write from the depth of my heart.

Just a few short decades ago, with the careless wave of a finger, the fate of thousands of humans—dehumanized cargo of unwanted Jews—arrived at a Nazi concentration camp to face what was called the Final Solution. The following narrative is condensed from Victor Frankl's personal account in his book *Man's Search for Meaning* in which he describes his experience on the day he first arrived. Frankl was an author, psychiatrist, and survivor of the infamous Nazi concentration camp in Poland called Auschwitz.

The architects of the Final Solution were not willing to wait to eradicate these people they deemed an inferior race by the slower process of forced sterilizations. They concocted a faster, more efficient way to purge the societal genes of these unfortunates they labeled inferior races of the Fatherland. Had the Germans won the war, they would have done likewise in the rest of the world.

In his biographical account of the terror he suffered, Frankl writes of the 1,500 Jews herded with him like cattle into railroad boxcars, with 80 people stuffed into each one. These trains, disguised with such deceitful and sinister names as the Charitable Transport Company for the Infirmed delivered the anguished, unsuspecting, condemned, and dehumanized cargo to their morbid destinations.

Everyone had to sit on their luggage, which held the last few remnants of any worldly possessions they might have had. Many suitcases contained family heirlooms and jewelry that had been passed down from generation to generation. Some contained memorabilia in the form of photographs and letters from loved ones that far surpassed in value any temporal wealth that gold or jewels could ever possess.

They had already lost their businesses and homes and whatever furnishings they might have had. All they and their ancestors had strived for in their lifetimes had been lost in a single stroke. But with tenacity and a little luck, they naïvely thought that perhaps they could manage to keep some of their family heirlooms and memorabilia.

The frightened families huddled together as they helplessly hoped for the impossible. Little did they know that even the clothes on their backs and the fillings in their teeth would be taken from them to aid the Nazi cause. Those sent to the gas chambers were used as raw material for the war effort. Soap and even lampshades were made from their remains.

For several days, they traveled without food, drink, or even hygiene, not knowing where they were going or even what to expect when they arrived at their undisclosed destination. Some hoped against hope that they would be taken to some factory to be used as forced slave labor. Few were.

Mothers clung to their children in quiet desperation, attempting to still the anguish of the uncertainty looming in their bosoms. The boxcars were so full that only the top sections of the windows let in any light.

After several days, the weary travelers finally reached their destination in the predawn hours. In the twilight of the morning, someone was able to read the ominous sign—Auschwitz—that appeared as the train slowed to a stop, and their hearts sank. As the gray light of early morning dawn broke, they scrambled to look out the windows, and there before them was an immense camp rimmed with barbed wire with sentry posts gradually materializing out of the semi-darkness.

Shouts and commands were given from the outside while guard dogs barked incessantly before the boxcar doors swung open and the guards commanded the weary travelers to disembark. They were greeted by a specially chosen group of inmates who, unbeknownst to them, had been well fed and trained for this very purpose. They were fellow Jews whose jobs were to receive the unfortunate new prisoners.

These trained Jews greeted the new arrivals with laughter and round rosy cheeks, giving the newcomers the hoped-for illusion that perhaps things would not be so bad and they might, after all, be well taken care of. But such was not the case. It was a pragmatic, heartless, and calloused ruse to make them cooperative and docile.

These specially kept prisoners were charged with relieving the newcomers of their luggage and everything in them in an orderly manner. The Jewish inmates were fluent in every European language, and that is why they were chosen and allowed to live. The heartless ruse was simply a pragmatic way of keeping the unsuspecting dehumanized cargo quiet and cooperative in the morbid process that was to ensue.

Eventually, all 1,500 of them were herded into and held for some time in a wooden shack built to accommodate perhaps 200 people. When the German officer in charge arrived, the guards formed them into two lines for inspection—women and children on one side and the men on the other.

The lines then filed past the SS officer in command as he coldly and calculatingly inspected each person carefully. Frankl describes this surreal scene in his own words as he was forced to file past the camp SS commandant:

> Then I was face to face with him. He was a tall man who looked slim and fit in his spotless uniform. What a contrast to us, who were untidy and grimy after our long journey! He had assumed an attitude of careless ease, supporting his

right elbow with his left hand. His right hand was lifted, and with the forefinger of that hand he pointed very leisurely to the right or to the left. None of us had the slightest idea of the sinister meaning behind that little movement of a man's finger, pointing now to the right and now to the left, but far more frequently to the left.

It was my turn. Somebody whispered to me that to be sent to the right side would mean work, the way to the left being for the sick and those incapable of work, who would be sent to a special camp. I just waited for things to take their course, the first of many such times to come. My haversack weighed me down a bit to the left, but I made an effort to walk upright. The SS man looked me over, appeared to hesitate, then put both his hands on my shoulders. I tried very hard to look smart, and he turned my shoulders very slowly until I faced right, and I moved over to that side.

The significance of the finger game was explained to us in the evening. It was the first selection, the first verdict made on our existence or non-existence. For the great majority of our transport, about 90 percent, it meant death. Their sentence was carried out within the next few hours. Those who were sent to the left marched from the station straight to the crematorium. This building, as I was told by someone who worked there, had the word "bath" written over its doors in several European languages. On entering, each prisoner was handed a piece of soap, and then—but mercifully I do not have to describe the events which followed. Many accounts have been written about this horror. We who were saved, the minority of our transport, found out the truth in the evening. I inquired from prisoners who had been there for some time where my colleague and friend P——had been sent.

"Was he sent to the left side?"

"Yes," I replied.

"Then you can see him there," I was told.

"Where?" A hand pointed to the chimney a few hundred yards off, which was sending a column of flame up into the grey sky of Poland. It dissolved into a sinister cloud of smoke.

"That's where your friend is, floating up to Heaven," was the answer. But I still did not understand until the truth was explained to me in plain words.[11]

For the lucky few, this was but the beginning of a long ghastly journey fraught with great sadness, suffering, and unimaginable pain—pain that words could never hope to adequately describe to those of us who have not experienced it. Pain that endured unrelentingly for years and which neither the reader nor this author could ever adequately empathize with. Pain created by human beings on other human beings. Cruelty that is scarcely found in nature, for even the wild beasts almost always only kill to eat.

What could cause people to harden themselves to such an extent that they would treat other human beings with such calloused disregard? What could blacken people's hearts and make them so insensitive to the sufferings of fellow human beings? I will tell you. It is the devaluation of humans from the rightful and lofty place given to them in the scriptures as the bearers of the image of God.

It is when humans deny the existence of God and adopt, instead, the naturalistic worldview that looks upon human life in a purely utilitarian fashion that their humanity is reduced to the level of a mere commodity. Take heed, my friend, for the horrors faced by these unfortunates may visit us again if we do not take care to understand the root of them.

Think clearly and carefully. Do not allow yourself to drift into the wide and crowded road of the walking dead, dull of mind and bereft of wisdom. Consider the cost of intellectual and moral apathy.

If we are simply nothing more than an evolved animal and each individual human being is not crowned with an intrinsic worth of infinite value that transcends the rest of creation, then why not eliminate from the gene pool of our society those deemed inferior specimens by the elite in power?

I am certainly not insinuating that the Jews have defective genes. What I am saying is that under the atheistic, existential presupposition and since Hitler believed that they did, and since he was in power, then he cannot be condemned for his actions by any who espouse a naturalistic worldview.

But Germany did not arrive at this place in a vacuum. It was the existential philosophy accepted by the leaders of its government that allowed and even promoted such barbarity. It was the inescapable political application of this naturalistic worldview. This is the future that awaits us if we continue down the path our nation has taken.

The dreaded Nazi ovens

What Is the Matrix of Reality?

If we are simply evolved animals, then what is morally wrong with killing other animals? Without a supreme arbitrator of morality, the survival of the fittest is the inescapable and logical conclusion of this naturalistic worldview. When an animal of prey in the wilderness hunts and kills another animal, we don't arrest and prosecute the animal, nor do we condemn the animal as immoral and evil, because that animal is simply exercising its animal instinct. If humans are a product of evolution, then we must conclude that violence or cruelty is just an inherited animal instinct. Moreover, the rule of the day is the survival of the fittest. Simply put, the guy with the biggest stick is right.

This is precisely the philosophy Friedrich Nietzsche expounded, and it heavily influenced Hitler's thinking in his sociopolitical policies. And alarmingly, it is the same reprehensible philosophy currently being naïvely and uncritically accepted by our postmodern culture in America, and it is fueling the socialist-globalist agenda.

How can we be so naïve as to blindly hurl ourselves into the same dark abyss that will invariably lead to death, destruction, and untold misery for our fellow brothers and sisters? As I read the cruel tortures of these victims and consider the stories I have heard my father tell of his prison experience as Castro's political prisoner, I cannot help but be incensed at the utter apathy that permeates my culture in this regard. We have uncritically accepted the liberal humanistic philosophy that will inevitably lead to an inhumane tyrannical government in America.

***

Our culture has naïvely built an illusory moral structure on the firm foundation of an over-inflated helium balloon that is bound to eventually explode and bring everything crashing down upon itself. The acceptance of relativistic values is the opening of Pandora's box and the straight road to tyranny.

***

Nietzsche properly rejected as hypocritical the liberal humanistic philosophy, which, after abandoning the concept of God, tried to build a "moral" system for the functioning of society based on the memory of Judeo-Christian morals. He was, without question, completely right in rejecting this hypocritical attempt to remain where there was no rational basis from which to support such optimistic "moral" notions.

Liberal humanists' attempt to build a superstructure that could allow society to protect the freedom of the individual is absolutely incongruous with their fundamental naturalistic presupposition. But spurred by their human need for transcendence, liberal humanists attempt to build religious-sounding structures to alleviate their intrinsic need for personal meaning. Modern man has moved further down the line of despair to postmodern man.

These liberal, humanist, progressive idealists have proposed that faith is what matters and that the object of faith is, for all practical purposes, very inconsequential. They have attempted to establish a belief system without God as a basis for morality yet with the purpose of providing some positive moral crutch upon which to lean their lives and culture. Whether it is a pantheistic, occult, or postmodernist, relativistic, theological socialist construct is immaterial. They all end up in the same place because they begin with the same relativistic foundation.

Ironically, modern man accepted this relativistic philosophical position in order to escape the notion of a higher being who holds us accountable to an absolute standard of morality. But the sterility of this worldview has led modern man to a new place. Postmodern man has now attempted to revert to the use of "God words" to mask the sterility of that worldview.

They have opted to turn to a mystical faith while maintaining the very same relativistic foundation. There is no real movement away from the Hegelian foundation. There is only the disguise of the Hegelian worldview in religious terminology.

The brute fact is that our culture—in some cases naïvely and unsuspectingly in other cases with partial understanding, but in all cases uncritically—has accepted this relativistic foundational presupposition without completely understanding the horrendous implications of sanctioning such a destructive worldview.

In our rebellion toward God, we have developed a calloused and myopic way of thinking in order to rationalize our sins. And our self-delusion has brought us to the edge of a precipice in our culture. I fear that it has brought us beyond the edge and into the great crevasse that Solzhenitsyn so eloquently warned us about.

But not even the depth of the ravine has moved us from our self-imposed horse blinds. Most people are gladly jumping off the cliff in order to assuage some impulsive lust, while desperately attempting to escape the guilt associated with a moral standard dictated by a just and holy God. Instead, modern man has reveled in the ecstasy of the moment, unconcerned for the disastrous consequence at the end of the fall. Sadly, it is our progeny who will pay the lion's share for our indiscretions as we hurl them headlong into the disastrous consequences created by our rebellion.

Our culture has entered recklessly and full throttle into the Age of Vericide. Rebellion has been legitimized and cleverly masked under the label of independent thinking and nonconformity. And immorality has been condoned, accepted, and perhaps even preferred under the auspices of freedom of choice within a philosophical framework in which all choices are equally permissible.

But in the end, in the framework of real reality, all of it is nothing less than wickedness. Know this, my friend, the universal Judge will inevitably hold all wickedness to account. The Holy Scriptures unabashedly acclaim, "It is appointed for men to die once and after this comes judgment" (Heb. 9:27). There is no escape from justice no matter how much you convince your brain otherwise.

There is but one road out of the valley of judgment, which could bring us across the mountains of wickedness. There is only one avenue of escape: the redemption found only at the foot of the cross when we come to Him in faith with empty hands. God's timeless and immeasurable grace has opened the Way for all humankind.

Why did we arrive at this juncture in our culture? Because we have been incapable of or unwilling to see through the clever smokescreens used to justify and promote liberal humanists' atheistic worldview. We have allowed their illusory framework of situational ethics to replace the Judeo-Christian worldview that was responsible for giving us our freedoms. How can we stand by idly as the tyrants of humanity offer us the poisonous fruit of the Age of Vericide?

Nietzsche correctly rejected this illusory framework of "morality," pointing out that it was built in a vacuum. If there is no God, then there is absolutely no foundation from which any semblance of a moral construct can spring forth. If there is no God, then only cruelty and violence are the matrix of reality! But how did it come to this?

Notes

1. Aleksandr Solzhenitsyn, *Warning to the West* (New York: Ferrar, Straus and Giroux, 1976), 79.
2. Martin Luther, quoted in Francis Schaeffer, *The God Who Is There*, Volume 1 (Westchester, IL: Crossway Books, 1982), 11.
3. Francis Schaeffer, *The God Who Is There* (Downers Grove, IL: Intervarsity Press, 1968), 20.
4. Glenn Beck, *Liars: How Progressives Exploit Our Fears for Power and Control* (New York: Threshold Editions/Mercury Radio Arts, 2016), 20.
5. Saul Alinsky, *Rules for Radicals* (New York: Vintage Books, 1971), 10–11.
6. Ibid., 78.
7. Ibid., 116–118.
8. Ibid., 61.

9. Aleksandr Solzhenitsyn, *Warning to the West*, 16–17.

10. Adolph Hitler, quoted in Jacques Ellul, *Violence: Reflections from a Christian Perspective* (Eugene, OR: Wipf and Stack Publishers, 2017), 130.

11. Victor Frankl, *Man's Search for Meaning* (New York: Pocket Books, 1961), 30–31).

CHAPTER 3

● ● ●

FACING THE MINOTAUR – DES GEITES

How is it that the Hegelian, naturalistic paradigm has been allowed to become the foundational matrix of our postmodern civilization both in Europe and America? To understand this historical development, we must begin at the last paradigm shift at the end of the Medieval Age. It was at this juncture in our civilization that the seed of the present Hegelian paradigm shift was sown.

After the awakening of the Reformation and its atheistic counterpart, the Renaissance, modern man drifted in two very different directions. There was an acute hunger for knowledge and a reawakened interest in learning, which were displayed by both Renaissance and Reformation thinkers.

However, each philosophic direction began with a very different foundation and proceeded to a very different outcome. Understanding the differences between these two competing worldviews is essential to understanding where we are as a society today. And, more importantly, it warns us of where we are going.

The Reformation brought humans to an awakening built on the foundational presupposition that humans, created by a loving and personal God, have a basis from which to study this creation. For this reason, it assumed that true truth would never contradict God's truth.

In fact, as a result of this worldview, which categorically stipulated that we live in a designed and previsioned world, humans came to the logical conclusion that the universe is orderly. Consequently, when this worldview is properly and carefully studied, humans can come to understand these laws and the universe that God has charged us to protect and understand. We are not to exploit it but to better understand the mind of the creator Himself and to better our own lot.

Moreover, it was implicitly understood that because a personal creator created human beings, there is an intrinsic congruency, or harmony, between humanity and the rest of His creation, which we can come to learn through careful observation and study. The natural outworking of this philosophic position, then, is that because the universe has a design, it can be discovered and understood through reason and scientific methodology.

This was the intellectual foundation that birthed the scientific revolution that ensued. No other foundation could have produced the scientific revolution. In complete contradiction to what we are being taught in our educational systems, the Renaissance could not have birthed the scientific process. If all is the result of random ordering, then there is no basis from which to assume that the laws of science could exist.

The Renaissance also saw people begin searching for knowledge, but from an entirely different presupposition that eventually led to the nihilism of Nietzsche. There are two terms used to describe this second philosophical development: the Renaissance and the Enlightenment.

It strikes me that the word *renaissance* means rebirth, a term that is distinctly occult and found universally in all occult theology. But that is not all. The second term, *enlightenment*, is also universally part of occult nomenclature. It refers to the attainment of esoteric or hidden occult knowledge. So in essence, both terms used to describe the second philosophical development came from occult theology. Hmmm!

They began from a philosophical basis that denied the existence of God and optimistically hoped that, as knowledge accumulated, humankind would find a way to unify the particulars with the universals. It was the great quest to find within a philosophic framework a way to provide a coherent field of knowledge.

I realize that this may be a bit hard to process for those who are not familiar with philosophy. Let me rephrase it in more common terms. In other words, philosophers aspired to find a cohesive system of thought that would provide humankind individually (the particulars) and society at large (the universal) with a system that could be applied consistently while still negating the existence of God.

That is, they sought to provide a superstructure for society that would allow civilization to function without God at the center. By superstructure, I mean a framework from which laws and civic boundaries could be drawn to allow a society to function.

All philosophical structures aim to create a cohesive system of thought that can govern the needs of the individual in harmony with the needs of the collective. But outside of God, no such construct can exist that can form a just and rational system of thought. These human attempts to superficially prop up a society with a relativistic legal and moral superstructure all lead to disastrous political consequences. Either the collective suffers in a form of anarchism or the individual suffers in a form of the tyranny of the collective.

Anyone interested in following this historical process through the disciplines of art, philosophy, music, education, and theology should read Schaeffer's indispensable book *The God Who Is There*. In fact, it is my sincere hope that anyone who reads my books will be encouraged to read all of Schaeffer's books, for without hesitation, I consider him the most important author of the twentieth century.

The Renaissance free-thinkers attacked the Reformation thinkers, characterizing them as mystical buffoons. They criticized (sometimes rightfully so) the traditions of religion and its excesses and abuses. Faith was depicted as the anachronistic superstitions of the uninformed, and skepticism became regarded as the hallmark of reason. Those who adopted a naturalistic-atheistic metaphysical position were then branded as the enlightened. Reason and faith were considered antithetical terms.

The Unnatural Division between Faith and Reason

The rise of Darwinism then sealed this unnatural division between faith and reason created by the Renaissance thinkers. Many theologians unthinkingly accepted this absurd travesty of logic. Faith—biblical faith—has never, is not, and never will contradict sound reason. In the Dark Ages, the excesses of the Roman Catholic Church and the hypocrisy of many of its leaders undoubtedly added greatly to the antipathy toward God that many experienced.

Some theologians, in an attempt to bridge the church and the Renaissance thinkers, threw out the baby with the bath water and abandoned the very basis of Christian truth, swapping it for a superficial Christianized version of relativistic thought. This gave birth to the Neo-Orthodox and Postmodernist movements that have infiltrated Christianity and are now spreading throughout Christendom as a cancerous growth devouring the truth. Many evangelical Christians became enamored with this progressivist social gospel idea and greatly promoted it throughout our nation, beginning in the early part of the twentieth century.

The first step into this dark corridor came with Immanuel Kant. Approximately 400 years after the birth of the Renaissance, Kant wrote of the categorical imperative. In *Groundwork of the Metaphysics of Morals,* he delineated his famous categorical imperative as a means to authenticate morality without God: "Act only according to that maxim through which you can at the same time will that it become a universal law."[1]

In other words, if you think it is applicable to the general populace, then you can accept it as good for you—a sort of truth or morality by popular demand through pure reason and without the need of God to substantiate it. This was Kant's feeble attempt to provide a union between the particulars and the universal through the horizontal venue of pure human reason alone.

The significance of this categorical imperative was that it represented a secular system that humans could live by and that they claimed to arrive at only by pure reason without the need to be anchored to divine revelation. They began to promote the idea of a natural morality that needed no God. Ironically, the maxim, of course, no matter how hard they try to evade it, is simply a reworking or corruption of the New Testament's golden rule.

Their artificial attempt is incapable of accomplishing what they want. It does not tie the particulars with the universal. In order for the categorical imperative to work in a society under a relativistic system, everyone must accept it as their personal worldview. How, then, can it work if those who refuse to accept it as their norm simply take advantage of those who do? Must we then force these, against their personal choice, to accept the categorical imperative? Again, we are back to force as the only arbitrator and the final reality in such a naïve worldview!

Nietzsche, being at least more honest or true to the implications of his atheistic presupposition, naturally rejected this type of argumentation as an obvious rephrasing of the Judeo-Christian concept he so abhorred. He correctly reasoned that if a human is an evolved animal, then in order for evolution to triumph in the ascent of the human species, humans must not seek the welfare of all humanity. Humans must facilitate the survival of the fittest and remove those individuals from the human genetic pool who would pass on genetic weaknesses to our future progeny. Humans must not impede the survival of the fittest.

Hence, these Judeo-Christian ideals that he labeled *ascetic* were the poison that, he reasoned, kept humans from progressing along the evolutionary ladder to the "super man." This business of establishing categorical imperatives was simply blasphemy to the intellectually consistent naturalist or atheist. It was a mirror of the Judeo-Christian morals he so despised.

Nietzsche, in his *On the Genealogy of Morals*, wrote equally caustically of Christians, as well as those liberal humanist idealists who promote a positivistic faith in humans, with absolutely no rational foundation from which to build. By the fiat establishment of a maxim that curtailed the absolute and unfettered freedom of an individual to do anything he or she pleased under the auspice of promoting the welfare of others, these liberal-humanist idealists were, to Nietzsche's clear thinking, no different from Jews and Christians who promoted what he termed negative or ascetic rules for living morally.

He simply lumped them together and called them no-sayers. He chose this term because, to him, they were saying no to the human spirit, curtailing a person's absolute freedom and keeping him or her from doing as he or she pleases.

Nietzsche stated:

> We 'knowers' are positively mistrustful of any kind of believers: our mistrust has gradually trained us to conclude the opposite of what was formerly concluded: to infer a certain weakness in the possible proofs of what is believed, or even its implausibility whenever the strength of faith of it becomes prominent. Even we do not deny that faith brings salvation: *precisely for that reason* we deny that faith proves anything,—a strong faith which brings salvation is grounds for suspicion of the object of its faith, it does not establish truth, it establishes a certain probability—of *deception*. What now is the position in this case?—These 'no'-sayers and outsiders of today, those who are absolute in one thing, their demand for intellectual rigour [*Sawberkeit*], these hard, strict, abstinent, heroic minds who make up the glory

of our time, all these pale atheists, Antichrists, immoralists, nihilists, these sceptics, ephectics, *hectics* of the mind [*des Geites*] (they are one and all the latter in a certain sense), these last idealists of knowledge in whom, alone, intellectual conscience dwells and is embodied these days,—they believe they are all as liberated as possible from the ascetic ideal, these 'free, *very* free spirits': and yet I will tell them what they themselves can't see—because they are standing too close to themselves—. . . . If I am at all able to solve riddles, I wish to claim to do so with *this* pronouncement! . . . These are very far from being *free* spirits: *because they still believe in truth.* . . . When the Christian Crusaders in the East fell upon the invincible order of Assassins, the order of free spirits *par excellence*, the lowest rank of whom lived a life of obedience the like of which no monastic order has ever achieved, somehow or other they received an inkling of that symbol and watchword which was reserved for the highest rank alone as their *secretum*: 'nothing is true, everything is permitted' . . . Certainly *that was freedom* of the mind [*des Geistes*], *with that* the termination of the belief in truth was announced. . . . Has a European or a Christian free-thinker [*Freigeist*] ever strayed into this proposition and the labyrinth of its *consequences*? Has he ever got to know the Minotaur in this cave by *direct experience*? I doubt it, indeed I know otherwise: – nothing is stranger to these people who are absolute in one thing, these *so-called* 'free spirits,' than freedom and release in that sense.[2]

Nietzsche was right on the money. If there is no God, as he was convinced, then these liberal humanists (pale atheists, as he called them) needed to face the fact that the consequence of an atheistic presupposition is the total and utter death of truth and morality: "Nothing is true, everything is permitted."

Such morality is what he referred to as the "ascetic ideal." He considered this morality "the danger of all dangers" since it kept the human species from reaching its highest evolutionary potential by eliminating the weaker elements. It means "facing the Minotaur," the concept symbolized by the "order of the Assassins"; namely, that force and might are all there is to reality.

***

Nietzsche proposed the establishment of a sociopolitical system organized along a strictly hierarchical concept (order of rank) and regimented by force. By this, he didn't mean to enhance life for all humans but instead for the generation of a few elite: those who, in his mind, were strikingly superior humans. Of course, to him this meant the Aryan race, to which he conveniently belonged.

He viewed Judeo-Christian concepts of morality such as pity and self-sacrifice as the evil poison that changed what he deemed to be humans' first understanding of good. Good was what he termed the *noble ideals*; that is, the freedom to plunder and ravage, as exemplified by the savage Celtic tribes that, in his view, were perverted by the "weak ascetic ideals" of the Judeo-Christian morals they later adopted.

The acceptance by society of these ascetic ideals, he theorized, was a pathological transition that Western man had to undergo momentarily in order to once again find the innocence of his primal freedom to plunder, pillage, and rape—to allow the animal in him to range free instead of being tethered by the suffocating moral restraints of the moral virtues he labeled ascetic ideals.

What if a regressive trait that lurked in the 'good man,' likewise a danger, an enticement, a poison, a narcotic, so that the present lived at the expense of the future. Perhaps in more comfort and less danger, but also in a smaller minded manner. **So that morality itself were to blame**

if man, as species, never reached his highest potential power and splendor. So that morality itself was the danger of dangers (emphasis added).[3]

To Nietzsche, the fettering of the powerful Aryan overlords with the ascetic morals was the single reason that destroyed the evolutionary process of weeding out the weak from the genetic pool of society. This ascetic morality was, in his view, the responsible culprit that consequently kept the species from reaching its highest potential. It was the culprit that stopped the intended evolutionary path of humankind, and therefore, it is the "danger of dangers" to be avoided at all cost.

He understood the concept of good as the carrying out of authority by the noble Aryan race. It is therefore an imperative to the enhancement of the future evolution of the human race in the ongoing process of the betterment of humankind to remove from the genetic pool those who are inferior and would therefore reproduce inferior traits. Humans must be absolutely free of any moral restraints to achieve the human potential—to become the Super Man.

Thus, it became the natural consequence of such thinking that Hitler would seek to incise from the genetic pool those who could not contribute to the imagined and imposed idea of what is good for society. They would, therefore, naturally be an economic and political drain on or hindrance to the elite of such a society. It was this philosophy that guided the Nazi movement in Germany to its horrendous consequences.

After World War II, Boston psychiatrist Leo Alexander was the Chief U.S. Medical Consultant at the Nuremberg War Trials. He wrote in a paper titled "Medical Science under Dictatorship":

Medical science in Nazi Germany collaborated with this Hegelian trend particularly in the following enterprises: the mass extermination of the chronically sick in the interest of saving "useless" expenses to the community as a whole; the mass extermination of those considered socially disturbing or racially and ideologically unwanted; the individual, inconspicuous extermination of those considered disloyal within the ruling group; and the ruthless use of "human experimental material" for medico-military research. . . .

It started with the acceptance of the attitude basic in the euthanasia movement, that there is such a thing as life not worthy to be lived. . . . Many people, including some in the medical profession, had accepted these principles before Hitler came on the scene . . . Alexander says that Hitler exterminated 275,000 people 'in these killing centers.' Then he adds that those so killed were to be only 'the entering wedge of extermination . . .' The methods used, and the personnel trained in the killing centers for the chronically sick became the nucleus of much larger centers in the East, where the plan was to kill all Jews and Poles and to cut down the Russian population by 30,000,000. The first to be killed were the aged, the infirmed, the senile and mentally retarded and defective children. Eventually, as World War II approached, the doomed undesirables included epileptics, World War I amputees, children with badly mottled ears and even bed wetters.[4]

If we are to prevent the repetition of the Nazi horrors perpetrated on humanity, then it must be understood by our culture that anti-Semitism alone did not account for the Holocaust. As a matter of fact, it was not even the primary reason for the Holocaust. Anti-Semitism festered as a natural by-product of the foundational naturalistic worldview that held to the maxim of the survival of the fittest.

It was the Hegelian view of relative truth that opened the door philosophically for a diminished view of humans that inexorably

lent to the inhumane treatment of humanity. Infanticide, euthanasia, and genocide are all legitimate tools for the rulers of a society under this naturalistic worldview.

It should therefore not surprise us that the Nazi political agenda developed medical ethics that viewed euthanasia as a positive thing. It began in Germany with the publication of *The Release of the Destruction of Life Devoid of Value, Its Measure and Its Form* by Alfred Hoche and Karl Binding. In this landmark work, the concept of destroying human life deemed by the state as absolutely worthless was considered an essential and positive step for society.

In this discussion, prominent Darwinists set the stage. Ernst Haeckl argued that infanticide, the killing of a child after birth, should not be regarded as murder but rather as "a practice of advantage, both to the infants destroyed and to the community."[5]

> It started with the acceptance of the attitude, basic in the euthanasia movement, that there is such a thing as life not worthy to be lived. . . . Even before the Nazis took open charge in Germany, a propaganda barrage was directed against the traditional compassionate nineteenth-century attitudes toward the chronically ill, and for the adoption of a utilitarian, Hegelian point of view. Sterilization and euthanasia of persons with chronic mental illnesses was discussed at a meeting of Bavarian psychiatrists in 1931.[6]

It is no surprise, then, that the ideals in the Magna Carta and the Declaration of Independence were, in Nietzsche's estimation, huge steps backward in the evolution of the human race. These magnificent political achievements, which owe their origins not to the Enlightenment but to the Reformation, represented to him the essence of evil because they were built on the Judeo-Christian concept of absolute truths that attempted to ensure equal justice for all members of a society. That is the fundamental difference between the theistic Reformation worldview and the atheistic Enlightenment worldview.

The natural outworking of the Judeo-Christian worldview rationally provides a basis for the assurance of equal justice to all—to the strong and to the weak. That, to Nietzsche, was the most heinous of crimes against the evolution of humankind. It was the application of this philosophical evolutionary construct of truth that brought upon us the horrors of the Nazi Party and the communist regimes of the nineteenth century. It is now the monster lurking behind the socialist agenda in our nation.

Moreover, the popularity of his worldview affected all of Europe for decades after. We cannot overemphasize the many social problems that were promoted by this worldview. The Eurocentric-colonialist mentality that it fostered created untold barbarities and injustices in the developing nations of the world that could not at the time resist the higher technological advantages of the European white person. And it is the same evolutionary ideology that framed the Hegelian-Marxist ideology that brought us collectivism in three poisonous flavors: progressivism, socialism, and communism.

But what is truly alarming is that our own society has now moved away from the Judeo-Christian bedrock upon which our nation was founded and without which there can be no individual or social justice. We are following in the footsteps of Hitler, Lenin, Stalin, and Mao, and the credulous and uninformed masses of our indoctrinated youth are unsuspectingly moving toward the very philosophical premise that justified the barbaric Nazi and Japanese ovens as well as the Soviet barges and the Maoist purges.

This did not happen overnight. It has been quietly brewing since the early twentieth century. We can see the roots of it in the eugenics movement that signaled the beginning of progressivism becoming accepted by the financial elite and academicians in America. The American eugenics movement that Margaret Sanger promoted is, in essence, the policy followed by the organization she founded to implement it into social action. That organization is called Planned Parenthood. It adheres to the same utilitarian philosophy, albeit in a more cleverly disguised form because it must in order to compete for our tax dollars. But before we delve into the history of the progressive movement in America, let us dig more deeply into the past—to the progressive movement's true roots in the Enlightenment era.

Renaissance vs. Reformation

And so we see that in Europe, there were two competing views that stemmed from the darkness of the Medieval Age: the Reformation and the Renaissance. They have also vied for supremacy in America, even from the beginning. Some of our early forefathers were deists, but the vast majority of them were solid Christians who understood well the historical precedent they were making in establishing the great American experiment in the stream of the Judeo-Christian worldview.

The Reformation principles were supreme during our early years as a nation since our demographics were overwhelmingly composed of evangelical Christians who had fled Europe to exercise their religious freedom. But since the twentieth century, the Renaissance ideology has taken the lead in our nation through the leadership of academia and government. The Judeo-Christian worldview has been relegated to the background of our society and banned from the educational system and many government buildings.

This happened as a result of two major factors. The first and least harmful of these is immigration. During the great influx of the nineteenth and twentieth centuries, most immigrants who came to America were non-evangelicals.

Through the great hall in the immigrant receiving center on New York's Ellis Island, opened in 1892, streamed in the next three decades almost four million Italian Catholics; half a million Orthodox Greeks; half a million Catholic Hungarians; nearly a million and a half Catholic Poles; more than two million Jews, largely from Russian-controlled Poland, Ukraine, and Lithuania; half a million Slovaks, mostly Catholic; millions of other eastern Slavs from Byelorussia, Ruthenia, and Russia, mostly Orthodox; more millions of southern Slavs, a mix of Catholic, Orthodox, Muslim and Jew, from Rumania, Croatia, Serbia, Bulgaria, and Montenegro. The waves of arrivals after the turn of the century were so enormous that of the 123 million Americans recorded in the census of 1930, one in ten was foreign born, and an additional 20 percent had at least one parent born abroad.[7]

This had some impact but would not have caused much of a problem by itself since, for the most part, the majority of them eventually assimilated into the American culture while keeping many of their ethnic distinctions. These by and large were amenable to the Judeo-Christian worldview. Although they were not evangelicals, they nevertheless embraced the principles of the U.S. Constitution and did not seek to undermine it from within, as is characteristic of the Islamic religion.

The second and most important reason is the spread of the Enlightenment ideology for the last century through the public educational system and the higher institutions of learning. This has heavily impacted our nation, robbing our young, pliable minds and filling our government with men and women who have rejected the Judeo-Christian worldview.

The rise of Darwinian evolution and the swift takeover of our higher institutions have promoted the Hegelian-naturalist ideology in our nation almost monolithically for more than 100 years. That is almost half our nation's age. Hegelian-naturalists' militancy has now radically increased as their numbers have grown. They now occupy key positions within our government,

which has given them the power to push their views, exclusive of all others. The Democratic Party is now, for all intents and purposes, basically the Socialist Party. We must recognize that there is no real movement toward plurality in their minds. There can be no syncretic union between these two opposing worldviews. There is in the very foundational presuppositions between the Reformation and the Renaissance a vast gulf that cannot be bridged.

The two begin from completely opposite worldviews and therefore produce two completely separate forms of societies. Sadly, they understand this better than Jews and Christians and have adeptly out-maneuvered us in the political and educational arena.

We have failed to understand the deep implication of adopting this relativistic worldview in our nation. The comparison of the American Revolution and the ill-fated French Revolution best exemplifies the difference in the respective consequences between the presupposition of the Reformation and that of the Renaissance.

Both began for the same reason. The French Revolution was a move toward the emancipation of the commoners from the stranglehold of the French aristocracy. The American Revolution was a move toward the emancipation of the commoners from the stranglehold of the British aristocracy.

Ironically, these aristocrats were the very same nobles Nietzsche referred to so idealistically. But the French Revolution was spearheaded by those who intended to build a society upon a platform of laws derived from a relativistic system. These were the "pale atheists" Nietzsche referred to so vehemently. And it is a matter of historical record that the French Revolution was largely engineered by the Jacobin occult secret society that is linked to Freemasonry.

Prior to the French Revolution, France was governed by a monarchy headed by King Louis XVI. His ostentatious living and disregard for the welfare of his people earmarked the inevitable end result of an elite class living off the backs of the oppressed working people. An eventual popular uprising against such unjust tyranny and economic repression is always inevitable.

But to better understand what happened, we must understand the events that led up to the French Revolution and how it influenced the outcome. During King Louis XIV's long reign, the Protestant movement received some severe setbacks in France, as Roman Catholic Cardinal Jules Mazarin effectively manipulated the French monarch behind the scenes.

> For most of his first three decades as monarch, Louis XIV (king: 1643–1715) was dominated by Cardinal Mazarin; during most of the later years he was totally under the influence of his favorite mistress, Madam de Maintenon, a devout Catholic. His constant wars over a long reign not only weakened France economically but in spite of early victories ended in the loss of major objectives. The Netherlands achieved total independence. The Protestant succession in England was recognized, and the Stuart pretender banished from France. France lost Nova Scotia, Newfoundland and Hudson Bay territory, retaining only Quebec in the New World. . . . The revocation of the Edict of Nantes (1685) was but one of many expensive blunders by a bad king, whose disastrous reign prepared his country for revolution.[8]

Not all members of the nobility were cruel taskmasters frolicking in licentiousness without regard for their constituents. Some of the members of this noble class were Protestant families who were greatly influenced by the austere teachings of John Calvin. They represented a considerable number of influential people who were well educated and did not share in the atheistic, occult, Enlightenment worldview of the Jacobins. But unfortunately, the other members of the nobility politically outmaneuvered them and ousted them from the land in order to minimize opposition to their planned takeover of the French monarchy.

Many of these Huguenot refugees, a result of the revocation of the Edict of Nantes, had been members of the nobility and, as such, were well educated persons of substance. They made a significant contribution to the development of the principles of the concept of a Christian republic.

Since this evangelical Christian element in France was banished from the country, the French Revolution was alternately based on a humanistic foundation rather than the Judeo-Christian worldview. When we compare this to the American Revolution, the end result of the two revolutions speaks for itself.

The American Revolution led to a successful government with a representative form of a republic that strove (although imperfectly) to assure equal justice and freedom for all members of its society, regardless of their financial status. On the other hand, the French Revolution led to an anarchistic disaster that filled the streets of France with human blood from guillotines and eventually and inevitably led to a totalitarian dictatorship in order to quell the chaos. Why?

Under their humanistic and relativistic presupposition, there was no absolute base on which to pin or sustain any law as absolute. The relativistic underpinning consequently led to mob rule and inevitable complete anarchy. Hence, the immorality and utter cruelty that followed was unchecked by any possible restraints. The guillotines killed thousands of innocents as well as the aristocracy and nobility who had previously callously disregarded the welfare of the commoners.

Men, women, children, and anyone who may have been even minutely associated with them (cooks, gardeners, and servants) were senselessly beheaded. The resulting state of affairs created such rampant violence and injustice that in due course, the people longed for the days of Louis XVI.

This inevitable anarchy brought more misery and less justice than when they were under the reign of their former insensitive, insensible, and incapable king. And the chaos that followed made the country ripe for the inevitable takeover of a totalitarian dictator such as Napoleon, who promised order.

Because of the revocation of the Edict of Nantes, France had expulsed all Protestant families and had therefore depleted many of the finest and most educated minds that could have influenced the revolution toward a different outcome. The French Revolution, based on Enlightenment ideals, failed miserably to produce equality, justice, and fraternity. It led back to an autocratic government.

But although the previous revocation of the Edict of Nantes led to the displacement of 50,000 Christian families, which created untold hardships on them and kept France from becoming a Protestant nation, there is a silver lining to the cloud. The nations of the Netherlands, England, Switzerland, the Protestant German states, and Prussia all accepted the French Huguenot refugees, and their presence in these countries significantly influenced them toward republican ideals espoused by evangelical Christian thinkers.

What Satan meant for evil, God used for good, as many of these countries later followed, although in a diminished sense, in the footsteps of the American experiment. The American form of government is historically unique, rooted in the absolute principles of basic human rights espoused by the Christian Reformers who championed the notion of equal justice for all members of society, regardless of social status, nationality, intelligence, or religious background.

However, to be completely candid, the Reformation was far from perfect, as any human endeavor is always plagued by our fallen nature. The Reformation itself was a process of gradual enlightenment. As the successive generations were able to peer more deeply into the Word of God, they consequently shed the trappings, accruements, and acculturations that had been added to the church since it had been legalized in the Roman Empire.

To the extent they were able to accept the precepts of the Word of God, they were able to put into practice a more just form of government for the people. These accruements and empty rituals are the inevitable march of all religions as people take the truth of God and envelop it with their own vain attempts to improve God's eternal and timeless truth.

The early Reformers were able to see a great deal of the reforms that were necessary to bring the church back in line with the Holy Scriptures, but they were also human and fallible. Step by step, the Holy Spirit of God began to illumine the following generations of Reformers, and layer after layer of Satan's evil doctrines that had blinded and enslaved the church began to be lifted. This was the great accomplishment of the Christian Reclamation, as I refer to the Reformation.

> It got rid of the encrustations that had been added to the Judeo-Christian world view and clarified the point of authority—with authority resting in the Scripture rather than church and Scripture, or state *and* Scripture. This not only had meaning in regard to doctrine but clarified the base for law.[9]

The Reformation thinkers understood that true truth, as Schaeffer exclaimed, is based on the propositional revelation of God. Therefore, morality is not subject to a popularity vote. That is what sets us apart and above the democratic system of the Greek city-states.

Modern textbooks are absolutely wrong in saying that our form of government was based on the Greek city-states. It was based on the Reformation ideals garnered from the Hebrew Scripture. The legitimacy of a law is not based on popular vote. It is based on the absolute bedrock of revealed propositional truth. There was no *Lex Rex* in the Greek city-states. It was the dictatorship of the majority.

Absolute moral imperatives are based on the revealed mind of God, and thus there is a basis for defending the absolute and infinite value of life for every single individual in this world, regardless of their social status, age, gender, race, intellectual prowess, or physical abilities.

Our nation's form of government is not a pure democracy, as we will soon discuss.

The Inevitable March of Nations

It is sad that in Germany, perhaps the single most influential nation of the Reformation, the ideals of the Christian Reclamation were soon forgotten as religion once again began its inevitable human process of corruption. It is fair to say that Satan was extremely wroth and incensed with great fury at Germany for her pivotal role in the setback he suffered in his historical quest for global domination.

No doubt he worked feverishly to corrupt her and destroy the moral stature of this great nation in an attempt to reestablish his march toward a global kingdom. For the same reason, he had lowered all guns, working feverishly to corrupt Israel. Today, he is doing the same with America. It is the inevitable march of nations.

In just 400 years, Germany went from Luther to Hitler. This is hard to understand, and yet it is the very norm of human nature and the reality of living in a fallen world. The once truly progressive (in the right sense of the word) and truly enlightened (also in the right sense of the word) Protestant religions began to sink into the gradual and inevitable quagmire of tradition. The vibrant church once again became encumbered with extraneous rituals and traditions.

The vitality of the Christian church eventually diminished to such an extent that it not only allowed the atheistic concepts of Nietzsche and his like to take over the nation, but there was also a significant rise in occultism. Such is the history of all humanity throughout all ages and without exception.

History has unquestionably documented that the brilliant insights of the Reformation thinkers were eventually overshadowed in Europe by the darkened ideals of the Renaissance. Christianity failed miserably to contend in the free market of ideas. Christians retreated to their churches and allowed the atheistic Enlightenment ideology to take over their culture. I love to travel to Europe and marvel at the architecture of their churches. But it is so depressing that they are now only mausoleums of the Christianity that was. America is now following that same path.

Today, we find ourselves living in the unfortunate eclipse of the ideals of the Christian Reclamation (Reformation), and the paradigm shift toward the Renaissance ideology is fast upon us, even in America. The Age of Vericide is in full bloom, releasing its noxious perfume with venomous consequences upon unsuspecting and ignorant admirers of its artificial beauty. It is the siren song that lures us into certain shipwreck and death.

For some time, the two ideals coexisted side by side, and nations vacillated back and forth as people championing different ideals took control at one moment or another. But every day the sun sets, the spider weaves its web ever tighter around our planet. Today, the nations continue to move further away from the principles established upon the revealed propositional truth of the scriptures.

It can be plainly stated that Europe has completely shifted from the Reformation ideals toward the occult Enlightenment ideals that more easily facilitate the unification of the nations into a global kingdom. The European socialist union may have coalesced only recently, but the shift that made it possible took place more than a century ago. And we must take note that the result of that shift has left its indelible mark on the world through the countless victims of two world wars.

But we must not give the impression either that at one time the European nations were wholly following the Reformation ideals. Immediately after the Christian Reclamation, most European governments did not completely adopt the principles of the Judeo-Christian worldview. There was, in every case, only a superficial Christian veneer.

Most of the time, the mandates of the governments were predicated on pragmatic and economic interests devoid of any form of higher morals by the elite in power. The sad truth is that even the superficial veneer of Christianity did not last long, and within 200 to 300 years, it has deteriorated significantly into our present dilapidated state of affairs.

Even within those nations that did not immediately deteriorate after the awakening of the Christian Reclamation in their Christian worldview, the full acceptance of the true and radical implications of its ideals was painfully slow. For example, the separation between the Anglican Church and the Roman Catholic Church was one of political convenience and had little to do with a genuine desire to return to the orthodox doctrine of the early church, at least at the beginning.

Our Nation's Roots

The Anglican Church in England was still steeped in many Catholic trappings, which frustrated the Pilgrims who wanted to return to the biblical model of the church. Many evangelicals in England had become impatient with the slow rate of reform in the Anglican Church. And, like the Huguenots in France, the Puritans and Pilgrims desperately sought to disencumber the church of any of the human trappings that embellished and obscured the truth; that is, the traditions that had been added to Christianity throughout many years since the time of the early church fathers, which disavowed the gospel of salvation by faith through grace and apart from our human efforts (Eph. 2:8–9).

The Pilgrims, however, differed from the Puritans who were committed to purifying the church slowly from within in a less radical form. As a result, the crown of England as a whole more or less tolerated the Puritans, although they frowned upon them.

The Pilgrims, on the other hand, were considered much more dangerous and radical, for they maintained that the Anglican Church was corrupt beyond reform. They would swear religious allegiance to neither king nor queen nor bishop, correctly claiming that Christ was the only head of the church. This brave stand cost them dearly.

In 1603, James Stuart, the king of Scotland, succeeded to the throne of England as James I. Since he came from Scotland, a predominantly Calvinistic rather than Anglican country, the Puritans and Pilgrims hoped he would grant them the freedom to travel faster down the path of reform. But apparently, James I was more interested in the power the Anglican Church gave him as head of the church.

They immediately presented him with a Millenary Petition of the Puritans, hoping he would support them and grant them their religious freedom. The king promptly called the Episcopal (Anglican) and Puritan leaders to Hampton in order to hash out the differences. But there, he gave his support to the more politically powerful Anglican Church, crushing the hopes of the Puritans and the Pilgrims.

Their hopes were cruelly dashed. The Pilgrim separatists led by William Bradford became the most persecuted religious group in England under the very same king they had earlier hoped would free them from the persecution by the Anglicans.

In retrospect, it can be said that at least one positive thing came out of this sad turn of events. The English language had changed so radically since the Tyndale Bible was translated that most people could not understand it. A commission was appointed as a result of this meeting in Hampton, which later produced the King James Version of the Bible. It became the most popular Bible translation in the English language until recent times.

As a consequence of their increased persecution, many decided to flee to Holland where they could enjoy greater tolerance. Unfortunately, King James had not only made it illegal for them to teach the truth but had also made it illegal for them to leave England, creating for them a desperate situation.

They were left with the only alternative: to practice civil disobedience or lose their religious liberty. In their first attempt to flee unnoticed, they were betrayed, captured, and subsequently thrown in jail for some time.

In their second attempt the following year, the men had already boarded the ship when the king's men sequestered their wives and children. The captain of the ship, fearing that he would also be incarcerated, subsequently did not allow the men to disembark. And so they sailed to Holland without their children and wives.

Fortunately, their families were able to join them later in Holland. However, their stay in Holland was a mixed bag. Although they enjoyed freedom, after a period of time, their children began to be corrupted by the influence of the morally lax lifestyle of the Dutch. As a consequence, the spiritual leaders of the Pilgrims felt led by the Lord to undergo the dangerous journey over the vast Atlantic Ocean to migrate to America where they could worship free from coercion and influence.

Such long and arduous voyages had never before been attempted by entire families. The severe hardships they encountered in the dangerous crossing, the clearing of the land, and the future prospects of dealing with the hostile natives must have been daunting for the hardiest of people. That they would attempt this with women and children was unthinkable at the time. This was indeed a revolutionary concept in their day.

In 1619, they boarded the *Mayflower* and set sail for a new country where they hoped to take the gospel of Christ to the natives. Initially, the *Mayflower* charted a course for Virginia. But strong storm winds sent it off course, northward hundreds of miles, landing instead in Cape Cod. Providentially, they happened upon a tract of land that had been previously cleared by an Indian tribe that had, just four years prior, been wiped out by a plague.

The group landed in the very dead of winter, which by itself could have spelled complete disaster for the group. But God's remarkable providence brought them upon the only tract of land for hundreds of miles that was not inhabited by hostile Indians but instead was already cleared and ready for farming. That specific region was an idyllic site, containing fertile land and an ample and suitable fresh water supply. Furthermore, had it not been for the kind intervention of the friendly natives, who also provided for them, they might not have survived that first winter.

Initially, they practiced a collectivist system of communalism that had been contractually imposed upon them by the merchants who had financed their trip. That is, no one owned private property; the entire community owned the land. It was, in essence, a mini-socialist form of government. But this communal system proved to be a dismal economic failure, and the colony was soon in grave danger of extinction due to famine.

Bradford realized that he must interject an incentive system if they were to survive. Going against the dictates of their financial backers, he assigned each family a plot of land to own and work. Each family consequently lived from the proceeds of their own labor. If they did not work their land, they would not have anyone else to blame for their hunger.

The system was a complete success, and from that point forward, there were no more food shortages. We must add that the right of private ownership is clearly documented in scripture. God divided the Promised Land between the 12 tribes of Israel. Each family was then given a specific territory. The ownership of the land returned to the families every 50 years, preventing the wealthy from hoarding the land.

The failure of the collectivist system has been shown throughout history as a dismal economic strategy. Bradford later wrote about this experience in no uncertain terms.

> The vanity and conceit of Plato and other ancients . . . that the taking away of property, and bringing [it] in community . . . would make them happy and flourishing; as if they were wiser than God. . . . [However, it] was found to breed much confusion and discontent, and retard much employment that would have been to their benefit and comfort.[10]

While the providential hand of God blessed the fledgling colony, back in England, repression and discontent continued to escalate. During the latter reign of James I, there arose a division between Parliament and the monarchy. The king's over-inflated ego had brought him into contention with Parliament as he pushed to gain absolute power for the monarchy.

By the time Charles I succeeded James I, there was a considerable rift between the monarchy and Parliament. Adding insult to injury, Charles appointed the See of Canterbury to an Anglican archbishop named William Laud, who used the Star Chamber as a tool to brutally suppress Puritans.

It must be stated that while the Puritans and Pilgrims were, in fact, separatists, they were not isolationists. They separated from society in their communities but actively participated in government and did their civic duties. It was these Christians who in their very first document, *The Mayflower Compact,* clearly stipulated the idea that rights come from God and not governments.

Their concepts of law and government did not arise from a vacuum. They reflected the thinking of Christians before them such as Henry de Bracton, an English judge who wrote *De Legibus et Consuetudinibus* around 1250 AD. In his work, he clearly delineated the Judeo-Christian worldview regarding law and authority. Standing against the norm of his day he stated:

> And that he [the King] ought to be under the law appears clearly in the analogy of Jesus Christ, whose vice-regent on earth he is, for though many ways were open to Him for His ineffable redemption of the human race, the true mercy of God chose this most powerful way to destroy the devil's work, he would not use the power of force but the reason of justice.[11]

We can also cite the writings of Samuel Rutherford, a Scottish Presbyterian minister (1600–1661) who authored *Lex Rex* in 1664.

This was a deeply revolutionary concept that refuted the absolute authority of kings. It was from Rutherford that John Locke borrowed and secularized this cornerstone of justice drawn from the Judeo-Christian worldview.

Christians are commanded by scripture to submit to the governing authorities. These authorities are instituted by God to protect the individual rights of their citizens, provide domestic tranquility, secure equal justice, and punish criminals with measures commensurate to their crimes. But when the governing authorities abandon their God-given sphere and take away the individual rights of their citizens, going against their God-given purpose, they abrogate their authority and become an illegitimate governmental structure that must be resisted.

This is not something new to the church. The apostles explicitly taught it. Even from the very beginning, Peter and John practiced civil disobedience when the law of the land strayed from their God-ordained limitations.

Peter and John were preaching to the people in Jerusalem. While in the portico of Solomon, they healed a lame beggar who for years had begged for alms by the Beautiful Gate of the Temple. Such was the power of their message that more than 5,000 people came to trust the Messiah as their Savior that day.

As the astonished people listened to Peter's impassioned teachings of Christ, Peter and John were arrested by the temple guards and jailed. The next day, the two apostles were brought to the Sadducees.

Now as they observed the confidence of Peter and John and understood that they were uneducated and untrained men, they were amazed, and began to recognize them as having been with Jesus.

They commanded them not to speak or teach at all in the name of Jesus. But Peter and John answered and said to them, "Whether it is right in the sight of God to give heed

to you rather than to God, you be the judge; for we cannot
stop speaking what we have seen and heard." When they had
threatened them further they let them go.

—Acts 4:13, 18–21

Two important truths may be gleaned from this powerful and important narrative. First, God's power is made perfect in the hearts of those who humble themselves before Him. He chooses the humble to oppose the proud, the weak to overcome the strong, the insignificant to confront the arrogant, the simple to confound the intellectually proud, the Davids to bring down the Goliaths.

Second, those who understand true truth must not ever allow any human or supernatural institution to usurp God's mandates. We are to obey God rather than people. But we must also count the cost. We must be willing to suffer the repercussions of taking a stand.

This example is repeated throughout the breadth and length of the body of Christ. Why do you think Christians were thrown to the lions during the three centuries of persecution by the Roman Empire?

The state unlawfully required that all its citizens must worship Caesar. From Christians' point of view, they could not comply without breaking the higher law of God that prohibits the worship of any other God than the creator.

From the Romans' point of view, Christians were disobeying civil law. The governing authorities viewed them as rebels who threatened the stability of the Roman Empire. The worship of Caesar was used to unite all conquered territories.

For this reason, Christians were thrown to the lions. The law of the land must not conflict with the higher law of God. If it does, then the law of the land ceases to be legitimate, and it becomes a tyrannical system that must be resisted. This has been the historical position of true Christianity from the very beginning.

We must be careful to note that resistance must always follow a course of action that exhausts all peaceful avenues of redress before escalating to force. Rutherford, drawing from the biblical example of David, explained that we must always choose to flee first before we come to the choice of violent confrontation. Nevertheless, at a certain point, it becomes legitimate to use force in order to oppose tyranny.

In the past, fleeing was a viable recourse to be taken before armed confrontation. For that reason, the Puritans fled to America rather than revert to violent resistance in England. Unfortunately, that may be a bit more complicated today in our modern world, for there is no place left to flee. America is the very last bastion of freedom in our modern world. It is the last trench. There is nowhere to go from here. It is here that I will make my trench and take my stand until my last breath.

Nevertheless, this fundamental ideology that guided and motivated our Founding Fathers is drawn directly from the Holy Scriptures, the very bedrock that brought forth the basic principles penned in the Declaration of Independence. Those principles did not come from Enlightenment thinkers as our modern textbooks incorrectly insinuate.

These Puritans and Pilgrims brought to America the idea that people had the right to govern themselves through a legal system that was embedded in the absolute truths expounded by the Holy Scriptures. It is only this foundation that can secure our fundamental rights as human beings.

In this respect, our very form of government owes its origin to the truths practiced by these brave Christians. The idea of our inalienable rights, of government by consent, of the separation of powers to forestall the centralization of power that is ever the fruit of the greed of fallen humans, and even the right of revolution stem directly from the Judeo-Christian worldview.

We are not speaking here of a theocracy. That must be clearly understood. The Bible specifically calls us to be functioning within the nations in order to affect the world with the positive influence of true truth. It does not call us to form a theocracy. It does not even call us to demand a state religion. That is another major difference between the Judeo-Christian worldview and Islam.

Rejecting the unbiblical merger of church and state practiced by both the Roman and Anglican Churches, the Founding Fathers sought to create a government based on the Judeo-Christian principles and yet not directly tied to a particular denomination. That, by the way, was the true intent of the First Amendment to our U.S. Constitution. The voice of the Puritans in England openly declared the illicit use of the state's power to cram down the throats of all its citizens a particular religious tradition.

The Anglican Church became more resentful and antipathetic toward these more radical factions of Christianity that sought to bring faith away from these acculturations and traditions that had been carried over from the Roman Catholic Church in order to align it with the teachings of the Holy Scriptures. The English secular power structure, which was well intertwined with the Anglican Church, therefore mercilessly persecuted these religious dissenters.

But it was their understanding of the scriptures that provided for America the very foundation of our Judeo-Christian republic. It is the concept of the inherent and inalienable rights of every human being because they equally share the semblance of their creator. And it is also responsible for the concept of a constitutional republic in which the citizens give the government consent to govern and where the absolute bedrock of law is above the government and popular opinion. This is absolutely crucial.

Moreover, it was not until the Puritans and the Pietists addressed the issue of slavery with clarity and conscience that the fundamental rights of every human being, regardless of their skin color or social strata, were proclaimed as an intrinsic and inalienable right of all humanity. Our modern historians who read history through their materialistic spectacles rarely acknowledge the role of the Christian church. That is an unmitigated travesty and the evidence of their subjective and militant atheistic bias.

It was their teachings, based on the Word of God, that brought abolition to the North in America and finally freed the slaves of the Antebellum South in 1863 and 1865. And it was also their leadership in England much earlier that caused the British to finally outlaw slavery throughout the empire in 1833.

Today, our nation is no longer guided by these fundamental Judeo-Christian principles that allowed for such freedom and liberty as never before experienced by any other nation since the inception of the nation of Israel. The shift in our government and especially in the courts is now almost monolithically slanted toward the naturalistic worldview. The unmitigated propaganda disseminated by modern textbooks in our public school system is nothing less than a rewriting of history to promote their naturalistic worldview. This is one of the major tools used to effect the shift in our nation's fundamental worldview, and it is working fabulously.

From the 1700s through the 1800s, the uniquely American spirit was characterized by a fiercely evangelical worldview that was highly skeptical of the federal government's encroachment on personal liberties. David Kennedy recounts how President Herbert Hoover wrote eloquently of this unique American Individualism:

> In Hoover's lexicon, the word that captured the essence of American Individualism was service. "The ideal of service," Hoover wrote in *American Individualism*, was a "great spiritual force poured out by our people as never before in the history of the world." It was a uniquely American ideal, and one that happily rendered unnecessary in America the repugnant growth of formal state power that afflicted other nations.
>
> Hoover revived, in a sense, the vision of a spontaneously mutualistic society inhabited by virtuous, public-spirited

citizens that had inspired the republican theorists of the American Revolutionary era.[12]

The fiercely independent American spirit of self-reliance and deep suspicion of central state power was a uniquely American experience that allowed our nation to spawn the largest middle class of any other nation on earth.

> Skepticism of government and of politicians who promised big things ran deep in the national DNA. People firmly believed in self-reliance, local control, and a strong civil society where neighbors volunteered to help another when things got bad. The federal government was a remote, abstract idea that never impinged on daily lives.[13]

Today, the federal government has grown to an over-bloated bureaucracy that is increasingly impinging on our daily lives and curtailing our cherished freedoms. Thomas Jefferson is turning in his grave. It all began with the advent of progressivism in America. The shift in America did not take place at the same time that it did in Europe. But it has taken place, nonetheless, only some 40 to 50 years behind continental Europe. Today, most Americans have accepted this shift without understanding its dire implications on our culture and our children's future.

The problem is that most Americans simply do not care enough to critically examine their worldviews. They simply adopt a worldview in much the same way we catch a cold, by the people we come in contact with. Americans have uncritically accepted the worldview stipulated by the slanted textbooks of our public schools, which do not offer education but rather indoctrination.

And yet our foundational worldview impacts and, more often than not, determines the position we adopt in the social issues of our age. It forms the foundation from which we can critically determine what is right and what is wrong. It determines how we view the issues of slavery, abortion, infanticide, euthanasia, cloning, embryonic stem cell research, transhumanism, socialism, supernationalism, collectivism, the family structure, anti-Semitism, radical Islam, globalism, and even our personal morality.

It determines whether we accept as correct the political ideologies and subsequent legal consequences resulting from counterfeit worldviews disseminated by such things as the religion of Islam or pantheism.

The Islamic and Progressive Threat to the U.S. Constitution

The reader must understand that all counterfeit worldviews lead to the same place, whether they espouse the syncretic concepts of the naturalist and pantheistic ideologies or the autocratic ideology of Islam. These counterfeit ideologies promise to be the most dangerous ideologies our nation and the rest of the world will face. The greatest danger to our freedom is the acceptance of the relativistic ideologies that are in complete antithesis to the Judeo-Christian ideology that fostered our constitutional form of government and our unique understanding of the value of every individual created in the image of God.

The pantheistic worldview erases the categories of good and bad in the same manner as the naturalistic-progressive worldview. They both deny the existence of absolutes, and therefore, rule becomes the arbitrary choice of the elite in power. Their favorite tool to bypass the will of the people then becomes the use of the Department of Justice (DOJ) mined by progressive ideologues to bypass the writ of the Constitution in order to force their relativistic progressive agendas upon we the people. Lately, the IRS, the FBI, and the DOJ have all been politicized to advance the progressivist-globalist ideology and circumvent the Constitution and the rights of the people. This is being orchestrated by a few political elites who believe their enlightened view gives them the authority to circumvent the Constitution and the will of the people in order to help the unenlightened masses.

Our threat comes both from within and from without. On the surface, Muslims claim to believe in absolutes, but in practice, that takes the form of an autocratic tyranny ruled by one all-powerful theocratic leader. Their doctrine, referred to as "the prophet of the day," is considered by their Sharia law to be capable of superseding previous views. Muhammad did this on several occasions when his sayings became a contradiction of a previous revelation. Thus, the words of theocrats become even more powerful than the words of the Koran. Moreover, their Islamic doctrine of progressive revelation and abrogation essentially makes their absolutes into relative choices. In the final analysis, there is little difference in the practical result of these counterfeit worldviews, which evidences their real author.

Muslims hold to a progressive view of revelation in which the old is abrogated by the new. To the Muslim, even the words of the Prophet Muhammad can be abrogated by a newer revelation. This is their doctrine of progressive revelation, otherwise known as the doctrine of the "prophet of the day." Muhammad himself taught this as new revelations from Allah contradicted previous revelations.

In fact, during his life, there were several instances when a new revelation abrogated the previous, or when verses were forgotten, and therefore Muhammad conveniently abrogated them.

> If we abrogate any verse or cause it to be forgotten, we will replace it by a better one or one similar. Do you not know that Allah has power over all things?
>
> —Surah 2:106

But, since "sometimes the revelation used to descend on the Prophet during the night and then he forgot it during the daytime," verse 2:106 was needed. Similarly troubling was the predicament of two men who had learned a Surah Muhammad had taught them, but then forgot it. Muhammad could not remember it either, but explained that this was because "it is one of those which have been abrogated, thus, forget about it." The scholars have little room for maneuver here: Allah changes his ordinances to fit the change of time and circumstances, although these remain apparently static to a mortal eye.

The fact that Muhammad got away with it all testifies to the faith or some other quality, of his faithful followers. The detractors claimed that "whenever he forgot what he related to his followers, he spared himself the embarrassment by claiming that God had abrogated what he conveyed to them before," the ever-faithful Aisha related how the Prophet heard a man reciting in the mosque and said, "May God have mercy on him, he has reminded me of such and such verses which I dropped from Surah so and so." The companions had to remind Muhammad of the forgotten verses at times, but when they were not available they had to be abrogated.[14]

How could Allah be the author of absolute truth if he allows truth to be negated and abrogated by another opposite concept? The mind of the real one true God the compassionate and merciful is not schizophrenic!

It is Satan who has invented the concept of the evolution of truth as one of his most effective tools to divert humans from the eternal and unchangeable truths of the true God. Of this one thing you can be sure. Any religion that tells you God's truths are not eternal comes from the human mind and the spirits of demons.

The learned men of Islam tell us they believe the psalms of David are inspired by God and are also, in fact, a transcript of the heavenly Book of Allah—in Hitti's own words, "an exact replica of a heavenly prototype."[15]

The psalms found in the Dead Sea Scrolls, which date to the second century before Christ (some 800 years before Muhammad) are exactly the same as the psalms that we have had in the Holy Scriptures from its inception for the last 3,000 years. There is no conspiracy to change the Holy Scripture. Christians and Jews did not take Muhammad out of the scriptures. He was never there.

Moreover, the narrative has not evolved and changed throughout the many centuries. This is evidence of divine protection. Not only are His words eternal, but He has also miraculously kept them from perversion throughout the many years. This is the true evidence of divine inspiration. And this is what God says about His absolute truths:

> *Praise the LORD, all nations; laud Him all peoples! For His lovingkindness is great toward us, and **the truth of the LORD is everlasting**. Praise the LORD! (emphasis added).*
>
> —Ps. 117

The learned men of Islam tell us that Jesus was a prophet of Allah. Hear the words of Jesus:

> *For truly I say to you, until heaven and earth pass away, **not the smallest letter or stroke shall pass away from the Law (Torah) until all is accomplished** (emphasis added).*
>
> —Matt. 5:18

There is not the smallest letter or stroke of the inspired scriptures that will ever be abrogated by any other concept, for they will all be without exception fulfilled.

Muhammad said he followed the God of Abraham and Moses. Moses said:

> *Give ear, O heavens, and let me speak;*
> *And let the earth hear the words of my mouth.*
> *"Let my teaching drop as the rain,*
> *My speech distill as the dew,*

As the droplets on the fresh grass
And as the showers on the herb.
For I proclaim the name of the LORD;
Ascribe greatness to our God!
The Rock! His work is perfect,
For all His ways are just;
A God of faithfulness and without injustice,
Righteous and upright is He" (emphasis added).

—Deut. 32:1–4

The one true God is the immovable rock whose work is *perfect*, and His truth needs not nor cannot be abrogated by anyone, for it is everlastingly true! It cannot be improved. It cannot be changed, and it will ever remain truth, absolute and eternal in nature forever.

To intimate that a new prophecy could abrogate an older prophecy is tantamount to accusing God of having imperfectly inspired the previous prophet. No—a thousand, thousand times no! "His work is perfect."

Muhammad's irrational justification ("Do you not know that Allah has power over all things?") is spurious logic. The idea that God can do anything is contrary to the very teachings of God. God cannot sin. The true God is not a finite being. He does not forget what He has inspired. His character and His truth are not subject to change. His truth is absolute and eternal. True truth does not evolve. Those who claim to speak for God and teach otherwise are usurpers and pretenders. The Allah of the Koran is not the God creator of the heavens and the earth. He is an impostor and usurper and quite likely a very powerful demon.

In the final analysis, we see that pantheism, Islam, and naturalism all extol a relativistic morality bent and stretched by the elite in power to suit their needs at the cost of the freedoms and liberties of the individuals over whom they rule. Without sound critical thinking, we become the mindless victims of misleading political ideologies with disastrous consequences to our future liberty.

The most dangerous aspect of the Islamic religion is its colonialist-globalist ideology. All Muslims who follow the teachings of the Koran, the Haddith, and the Jimma consider our Constitution an illegitimate form of government. Their call is to bring every nation under their Sharia law. Hence, they are forbidden to really assimilate into any other culture and to accept any other form of government as legitimate. They may feign assimilation at the beginning when their numbers are few. But as their numbers increase within the nations, they attempt to colonize, and their tactics become more violent and eventually brutal.

Our culture today is faced with a variety of virulent political ideologies competing for global supremacy, which run the spectrum from anarchy to supernationalism (e.g., Islam, white supremacists, Nazis) and collectivism. We must be able to understand the grave consequences of not knowing how to analyze each of these ideologies in order to understand how they will impact our future as individuals and as a society. Whether we like it or not, our ignorance, and or apathy, will also impact our future, but perhaps not in the way that will best benefit our children.

It is my prayer that as I venture into the heart of the social issues of great importance in our generation, the light of truth that comes from above will guard and guide your hearts and minds and show you that the road of faith and the road of reason converge upon the often lonely and narrow path of true truth.

May the immense depth of King David's simple psalm provide for you, as it has for me, the paragon to anchor truth in contrast to the ever-shifting sands of relativism blindly offered by our wilting world.

Thy word is a lamp unto my feet, and a light unto my path.
—Ps. 119:105 KJV

Notes

1. Immanuel Kant, *Groundwork of the Metaphysics of Morals* (Cambridge, UK: Cambridge University Press, 2012), xx.
2. Friedrich Nietzsche, *On the Genealogy of Morality*, ed. Keith Cosell-Pearson (Cambridge, UK: Cambridge University Press, 1994), 117–118.
3. Ibid., 8.
4. Leo Alexander, quoted in Francis A. Schaeffer, and C. Everett Koop, *Whatever Happened to the Human Race?* (New York: Crossway, 1976), 105–106.
5. Michael R. LaChat, "Utilitarian Reasoning in Nazi Medical Policy: Some Preliminary Investigations," *The Linacre Quarterly* 42: no. 1 (February 1975): 18.
6. Leo Alexander, "Medical Science under Dictatorship," *New England Journal of Medicine* 241 (July 14, 1949): 39–47.
7. David M. Kennedy, *The American People in the Great Depression: Freedom from Fear: Part 1* (Oxford, UK: Oxford University Press, 1999), 14.
8. Franklin Hamlin Littell, *The Macmillan Atlas History of Christianity* (New York: Macmillan Publishers, 1976), 105.
9. Francis Schaeffer, *A Christian Manifesto* (Westchester, IL: Crossway Books, 1981), 28.
10. Marshall Foster and Mary-Elaine Swanson, *The American Covenant, the Untold Story* (Thousand Oaks, CA: The Mayflower Institute, 1981), 83.
11. Henry de Bracton, *De Legibus et Consuetudinibus, Vol. 2* (Cambridge, MA: Harvard-Belknap, 1968), 33.
12. David M. Kennedy, *Freedom from Fear*: Part 1, 47.
13. Glenn Beck, *Liars: How Progressives Exploit Our Fears for Power and Control* (New York: Threshold Editions/Mercury Radio Arts, 2016), 26.
14. Serge Trifkovic, *The Sword of the Prophet* (Boston, MA: Regina Orthodox Press, 2002), 80–81.
15. Philip Kuhri Hitti, "The First Book," AramcoWorld, https://www.aramcoworld.com/en-US/Home.

CHAPTER 4

● ● ●

LAW, JUSTICE, AND FREEDOM

Modern man has fallen prey to the same temptation that Eve succumbed to in the Garden of Eden. It is the desire to "be like God," having the capacity to determine the difference between good and evil (Gen. 3:5). Once man accepts the philosophy that he can determine for himself, without utilizing God's revelation, the difference between good and evil, he starts believing he is autonomous from God. Then, man like a god, is separate and can work on his own to effectuate his purposes. Man can now proclaim with Nietzsche, "Dead are all Gods; now we desire the superman to live." The Christian religion is thus suspect, for it allows no supermen within its doctrines. It is a stranger in a world with human gods.[1]

The great deception of humanity, spawned at the very beginning of our race of beings, has become firmly imbedded in the very matrix of our human culture. This is the great lie that permeates society and deeply influences our thinking from childhood until death and ever perverts our perception of true reality. It is our most destructive inclination, which deceives us into thinking we can function in absolute autonomy from God.

Moreover, it evidences the very thing that we try most to ignore: our fallen state. It blinds us to "the enemy within," as Dr. Paul Rodriguez, former professor of psychiatry at Toccoa Falls College, so eloquently elucidates. It is the enemy within that arrogantly supposes that we can be our own gods. The essence of the great cosmic battle of the long defeat has been from the beginning the quest to usurp the throne of God. But we are not as those who have no hope. We know that the victory has already been won by our Savior.

The masks of this cosmic struggle are many, but their essence is ever the same. When humans turn from God, they essentially claim to be gods. They place themselves at the center of the universe. They make themselves the measure of all things.

We have already established that in such a reality there is no distinction between good and evil. Truths and morals in such a universe are but an illusion. And consequently, there can be no basis from which to establish any form of true justice. In such a universe, all is essentially sameness.

Only the Judeo-Christian concept of reality is able to provide a lucid continuity between unity and diversity. The universe is one because a personal God has created it, and all reality—the visible and the invisible—is therefore under His sovereignty. Moreover, we have each been created in His image and therefore possess unique qualities that provide for an infinite array of diversity and, most importantly, infinite value for each and every one of us.

It is because we have been created in His image that humanity is endowed with infinite worth and transcendental significance. The diversity has therefore a distinct function in His universal design. Each individual has unique significance and a special purpose for his or her life. This is big—very big! No other worldview outside the Judeo-Christian worldview provides for this confluence between unity and diversity. As Francis Schaeffer once told me, it's not that Christianity is the best answer; it's that Christianity is the only answer.

It is for that reason that the choices we make are significant because we can choose as individual beings to follow or rebel against His precepts and the express purpose for which we have been individually created. Consequently, when we choose to become our own gods, our choices have real and negative consequences. And conversely, when we choose to follow His precepts, our choices have real and positive consequences as we begin to fulfill His intended purpose for our lives.

If we deny the existence of God, we become victims of the very real and tangible cosmic battle that has raged all about us between the forces of good and the forces of evil from the very beginning of the human race. We can convince ourselves that these are simply fairy tales, but the impact will be real nonetheless.

The denial of God brings modern people to the doorstep of the false promise given to Adam and Eve by the serpent in the Garden of Eden. According to Dr. Rodriguez, it is "the false promise given to Adam and Eve in the Garden of Eden" that motivates humans to become enslaved, ever chasing the mirage of the self-healing and self-repair of the wound created in every soul by the space-time, historical severance from the spirit of humankind during the Fall.

It is that wound that creates in humans the many forms of pathological behavior that brings so much evil and suffering upon us. It is the very source of all the injustices that have plagued our human family from the time of Adam. It is the root of selfishness that leads to all other evils in our world.[2]

When we make our minds the center of the universe, we are expressing the most severe form of selfishness possible in our universe. This is at the root of everything that is evil, unjust, and perverse. And it is the reason that any sociopolitical ideology founded on an evolutionary or relativistic framework will always lead to the abuse of the common person by the elite in power.

Because we have been created in God's image, each and every human being is born with an innate knowledge that there is such a thing as good and that the opposite is evil. But without God's revealed propositional truth, we would be incapable of even knowing the essence of good and evil, for all human choices would then be on an equal footing.

Our human conscience tells us that there is good and evil, but without the revealed mind of God, there is no way to determine the specifics of those categories. For that reason, Satan always attacks the reliability of the propositional revelation of God. He attempts to discredit the Word of God in order to remove the foundation for divinely revealed true truth.

The postmodernist syncretic ideology that places all other religious books on the same level as the Bible is nothing more than a canon shot to the broadside of true truth. The scriptures alone carry the evidence of inspiration through the predictive prophecies that it contains. No other religion can claim otherwise.

Without the Holy Scriptures as the anchor point for true truth, there can be no reference point for morals and justice. Therefore, the idea that our faith should be removed and separated completely from our legal concepts, which are supposed to be designed to ensure justice, is completely irrational, absolutely absurd, and philosophically unjustifiable.

All legislation naturally presupposes a moral foundation from which it emanates. However, without an absolute standard from which to measure morality, all choices are essentially equal in value. In such a reality, there is no justification even for legislation since morality and justice are merely illusions. This is the stark outcome of an atheistic worldview. It is the only rational outcome of atheism. There is no *Lex Rex* possible when laws do not come from absolute morals. In a relativistic system, each person is *rex* over *lex*. And it is the foundation for the progressivist-socialist-collectivist view of a relativistic jurisprudence that serves only to justify their goal to attain political power at the cost of our personal liberties.

For this reason and as I have insisted before, the natural outcome of the naturalist worldview is always eventually tyranny. It is only a matter of time. Since there is no higher law, an elite few arbitrarily decide what is good for the people for that particular moment, thus binding the whole nation by their arbitrary choices. But you can be sure that what is good for the elite is always disguised in terms that supposedly benefit the collective.

Can One Legislate Morality?

Progressivists surreptitiously propagate the misconception that one cannot legislate morality. Moreover, they have elevated this false ideology into the pervasive dark echelons of political correctness within our society. But we must understand that this is simply an excuse to keep Christians and Jews from promoting their moral principles derived from the absolute standards revealed by God within the Holy Scriptures through legislation so their atheistic worldview can stand supreme and unopposed. It is the essence of tyranny.

The fact is that, contrary to their claims, all legislation is in its very essence the enforcement of a moral viewpoint. This modern, politically correct ideology is but another clever smokescreen designed to neutralize the Judeo-Christian worldview. It is impossible to create a law without some moral or ethical foundation that necessitates its existence.

If the authors of that particular piece of legislation do not reflect the moral principles of the Judeo-Christian community, then they will reflect whatever foundational worldview the legislators happen to hold, whether it is atheistic, pantheistic, Muslim, or occult.

Nevertheless, there are many who openly fume at the idea that Jews and Christians should involve themselves in political activities to promote the moral issues they consider to be of real importance in our culture, such as protection of the unborn. They create an artificial chasm between religion and politics and claim that never the two should meet.

The First Amendment to the Constitution is often cited as the legal basis for such a position. But this is a complete fabrication and contortion of the meaning of this valuable amendment. They completely disregard the intent of the framers and by fiat interpretation invent what is convenient to promote their naturalistic worldview.

This is the First Amendment to our Constitution:

> Congress shall make no law respecting an establishment of religion, or prohibiting the free exercise thereof; or abridging the freedom of speech, or of the press; or the right of the people peaceably to assemble, and to petition the government for a redress of grievances.[3]

There is a distinct difference between the government promoting a given religion or denomination and promoting the moral standards derived from the Judeo-Christian worldview. These basic standards encompass the framework necessary for the exercise of true justice in any government. The true and original intent of the framers of our Constitution when they created the First Amendment was never to divorce God from law. It was to keep any particular denomination from establishing preeminence over the others.

It had been the practice of European nations to adopt a single national denomination as their state church. This was what our Founding Fathers tried to avert. Each and every state in our fledgling union also adopted a single denomination as their state religion. That by itself shows that they never meant for us to separate God from government.

Our Founding Fathers wisely stipulated that the federal government should not do likewise. Moreover, they specifically stipulated that the federal government should not interfere with or prohibit the free exercise of religion.

In other words, the First Amendment prohibits them from segregating us and disavowing our right to represent our worldview in our democratic process, in our public educational system, and in the public square. That includes the right of Americans to speak their religious views and print them, to assemble peaceably to worship, and to petition the government when their rights are being infringed upon. It guarantees us the right to be able to vote according to our consciences. It certainly does not in any way stipulate that our laws ought not to be based on the ideals of the Judeo-Christian foundation for justice and equality.

These rights were absent in the European nations from which Americans migrated. Once a particular denomination became entangled with the secular power, there was persecution for all other denominations. For this reason, they fled Europe, and for this reason, the First Amendment was dear to their hearts. The modern misinterpretation is, in essence, the exact opposite of what the Founding Fathers intended. Their clever smokescreen has instead, against the very writ of the First Amendment, effectively interfered with the free practice of religion. Our Founding Fathers would turn in their graves to hear how their intentions have been twisted and profaned into an infringement on our ability to freely express our religious rights.

But that does not matter to those who are twisting it. The intention of our Founding Fathers is immaterial to a relativist. Their idea of law is not measured by case history but by fiat decision. It is an arbitrary and pragmatic decision made for a specific point in time as the elite arbitrarily and arrogantly decide what is good for the rest

of us. Their arrogance is superseded only by their condescending attitude toward others who do not share their "enlightened" view.

Our Founding Fathers sought to keep America free from the control of a given denomination or the humanly contrived organizations formed by people of any religion. But they well understood that the principles that comprised our particular form of government were derived strictly from the Judeo-Christian worldview. Jews and Christians who ignore this truth are negligent in their responsibility to continue in the light and path bravely undertaken by our forefathers.

Today, liberal judges in the U.S. Supreme Court who have reinterpreted our Constitution through their Hegelian magic spectacles have turned this fundamental meaning on its ear. They now seek to create an unnatural wedge between the government and God altogether.

The fact that our Congress should make no law that will establish a religion does not mean that religion should be absent from the government. If they interfere with our religious duty to vote our conscience and promote our moral worldview, then they are, in fact, prohibiting the free practice thereof and are therefore acting against the specific stipulations of the First Amendment.

Moreover, the First Amendment did not apply to state governments, but specifically to the United States Congress. Unfortunately, the U.S. Supreme Court many years later ruled that it applies also to the individual states, which were originally considered to be independently sovereign entities. Federalism on steroids has expanded illegitimately to infringe upon state rights. Globalists have consequently rammed their relativistic fiat judicial decisions down the throats of state governments when the original framers never intended this.

Our nation was founded as the union of free and independent states. It is no longer so. The expansion of the federal government has continuously eroded the sovereignty of the independent states as the demonic hierarchy continues to push toward the centralization of power that will eventually bring forth their long-planned global government.

The Separation Delusion – The Myth of the Separation of Church and State

Some 40 years ago, I was involved in the first National Evangelical Pro-Life Conference held in Ft. Lauderdale, Florida. My mentor, Francis Schaeffer, was the keynote speaker, along with many other notables of our time. One evening, about 12 of us young firebrands were invited to dinner on the second floor of a Ft. Lauderdale restaurant, which we have since affectionately referred to as the upper room dinner. Included in that group was the young Franky Schaeffer whom I still love and pray for, although he has now changed his views on some central social issues. Also attending was a young constitutional lawyer named John Whitehead who went on to start the Rutherford Institute, a legal practice that has helped many who have a Judeo-Christian worldview avoid becoming victims of our government's overreach and constitutional infringements.

A few years later, Whitehead wrote a book called *The Separation Illusion* in which he set a solid constitutional and historical case for the legitimacy of Christianity in the political arena under constitutional protection (I highly recommend it). Today, some 40 years later, I believe the progressivist agenda has become so advanced in the minds of the public that the separation illusion has become a concrete delusion.

If we could hear the talk of politically correct Democrats and even some RINOs (Republicans in name only), we would think the separation of church and state was chiseled in granite on the Washington Monument or the Jefferson Memorial. It is not. I have been to all the monuments in Washington, DC, and not a single one has that concept written in granite. Quite the opposite, the mention of God is chiseled into many of them.

The numerous times God was mentioned in our Declaration of

Independence is absolute proof that our Founding Fathers never intended to create an impassable wall of separation between God and government. That would have been adamantly refuted by any and all the signers of the U.S. Constitution, including the few deists whom the liberals are prone to worship.

For those who care to know, the writings of our forefathers leave us indisputable evidence that absolutely refutes this modern myth of the absolute wall of separation. We can cite as an example a proclamation by George Washington, our first president, by the specific request of both houses of Congress. Mind you, this proclamation had the backing of both the legislative branch and the executive branch of our government.

Washington understood well how rare and wonderful were the blessings of freedom we were given in this nation. He was the first to enshrine Thanksgiving as a national holiday. It was not to thank the Indians for helping the Pilgrims or to celebrate pumpkin patches and jack o-lanterns, but because he knew our federal government would not have been possible without God's providential care.

It has absolutely nothing to do with Pilgrims and Indians and turkeys. His official proclamation continues to remind us today that the providence of Almighty God was and is the source of all the freedoms we enjoy as American citizens. I challenge you to find any semblance of the modern liberal concept of the wall of separation between church and state in his speech:

> City of New York, October 3, 1789. Whereas it is the duty of all Nations to acknowledge the providence of Almighty God, to obey his will, to be grateful for his benefits, and humbly to implore his protection and favor, and Whereas both Houses of Congress have by their joint Committee requested me "to recommend to the People of the United States a day of public thanks-giving and prayer to be observed by acknowledging with grateful hearts the many

signal favors of Almighty God, especially by affording them an opportunity peaceably to establish a form of government for their safety and happiness."

Now therefore I do recommend and assign Thursday the 26th. day of November next to be devoted by the People of these States to the service of that great and glorious Being, who is the beneficent Author of all the good that was, that is, or that will be. That we may then all unite in rendering unto Him our sincere and humble thanks, for his kind care and protection of the People of this country previous to their becoming a Nation, for the signal and manifold mercies, and the favorable interpositions of his providence, which we experienced in the course and conclusion of the late war, for the great degree of tranquility, union, and plenty, which we have since enjoyed, for the peaceable and rational manner in which we have been enabled to establish constitutions of government for our safety and happiness, and particularly the national One now lately instituted, for the civil and religious liberty with which we are blessed, and the means we have of acquiring and diffusing useful knowledge and in general for all the great and various favors which he hath been pleased to confer upon us.

And also that we may then unite in most humbly offering our prayers and supplications to the great Lord and Ruler of Nations and beseech him to pardon our national and other transgressions, to enable us all, whether in public or private stations, to perform our several and relative duties properly and punctually, to render our national government a blessing to all the People, by constantly being a government of wise, just and constitutional laws, discreetly and faithfully executed and obeyed, to protect and guide all Sovereigns and Nations (especially such as have shown kindness unto us) and to bless them with good government, peace, and concord. To promote the knowledge and practice of true religion and virtue, and the increase of science among them and Us, and generally to grant unto all Mankind such a degree of temporal prosperity as he alone knows to be best.[4]

—George Washington

Apparently, George Washington would not make the cut in today's political arena in Washington, DC. The city that is named after him is so far removed from his beliefs that he would be ridiculed and ostracized. I dare say, he might be leading another revolution if he were to see what has been done with our nation.

How dare he suggest that the Congress of our federal government should "promote the knowledge and practice of true religion and virtue."[5] How naïve! Does he really believe that "it is the duty of all Nations to acknowledge the providence of Almighty God, to obey His will"[6]? Doesn't he know that there is an absolute wall of separation between government and God?

I am afraid that he would be tarred and feathered in the modern sophisticated cocktail parties of the elite politicos of our government for making such an obviously politically incorrect statement. Poor George! He would be the brunt of all the jokes in our "enlightened" modern media centers. He would be labeled an unsophisticated and archaic moralist.

Barack Obama, Hillary Clinton, and Bill Clinton would clink their martini glasses together and wink condescendingly. They would approach George, put their arms around him in a fatherly gesture, and gently pull him to the side in order to instruct him in the proper Washingtonian modern etiquette about the R word. They would patiently instruct him in their more "enlightened" interpretation of the meaning of the First Amendment and encourage him to keep religion and his personal beliefs to himself. Obama would probably pass him a joint and say, "Hey man, chill out with this God stuff."

For years our liberal indoctrination (public education) system has taught us that there is an absolute wall of separation between church and state. It has labeled Thomas Jefferson a deist and one who did not accept the entire scriptures as the Word of God, having abridged a book that contained only the sayings of Jesus from the Gospels. What they did not tell you is that he labored to do so in order to provide for the Native Americans the basic moral principles of the Judeo-Christian worldview. He feared that reading the many wars in the Old Testament would be counterproductive in maintaining peace with them.

Obviously, Jefferson did not abide by the modern misrepresentation of his faith and of this supposed absolute wall of separation erroneously extrapolated from his letter to the Danbury Baptist Church. We can clearly see in his Thanksgiving proclamation issued on November 11, 1779, when he was governor of Virginia, that he relied solely on "the glorious light of the gospel, whereby through the merits of our gracious Redeemer we may become the heirs of his eternal glory."[7] Below is the complete text of that November 11, 1779, proclamation appointing a day of thanksgiving and prayer:

> Whereas the Honourable the General Congress, impressed with a grateful sense of the goodness of Almighty God, in blessing the greater part of this extensive continent with plentiful harvests, crowning our arms with repeated successes, conducting us hitherto safely through the perils with which we have been encompassed and manifesting in multiplied instances his divine care of these infant states, hath thought proper by their act of the 20th day of October last, to recommend to the several states that Thursday the 9th of December next be appointed a day of publick and solemn thanksgiving and prayer, which act is in these words, to wit.

Whereas it becomes us humbly to approach the throne of Almighty God, with gratitude and praise, for the wonders which his goodness has wrought in conducting our forefathers to this western world; for his protection to them and to their posterity, amidst difficulties and dangers; for raising us their children from deep distress, to be numbered among the nations of the earth; and for arming the hands of just and mighty Princes in our deliverance; and especially for that he hath been pleased to grant us the enjoyment of health and so to order the revolving seasons, that the earth hath produced her increase in abundance, blessing the labours of the husbandman, and spreading plenty through the land; that he hath prospered our arms and those of our ally, been a shield to our troops in the hour of danger, pointed their swords to victory, and led them in triumph over the bulwarks of the foe; that he hath gone with those who went out into the wilderness against the savage tribes; that he hath stayed the hand of the spoiler, and turned back his meditated destruction; that he hath prospered our commerce, and given success to those who sought the enemy on the face of the deep; and above all, that he hath diffused the glorious light of the gospel, whereby, through the merits of our gracious Redeemer, we may become the heirs of his eternal glory. Therefore,

Resolved, that it be recommended to the several states to appoint THURSDAY the 9th of December next, to be a day of publick and solemn THANKSGIVING to Almighty God, for his mercies, and of PRAYER, for the continuance of his favour and protection to these United States; to beseech him that he would be graciously pleased to influence our publick Councils, and bless them with wisdom from on high, with unanimity, firmness and success; that he would go forth with our hosts and crown our arms with victory; that he would grant to his church, the plentiful effusions of divine grace, and pour out his holy spirit on all Ministers of the gospel; that he would bless and prosper the means of education, and spread the light of christian knowledge through the remotest corners of the earth; that he would smile upon the labours of his people, and cause the earth to bring forth her fruits in abundance, that we may with gratitude and gladness enjoy them; that he would take into his holy protection, our illustrious ally, give him victory over his enemies, and render him finally great, as the father of his people, and the protector of the rights of mankind; that he would graciously be pleased to turn the hearts of our enemies, and to dispence the blessings of peace to contending nations.

That he would in mercy look down upon us, pardon all our sins, and receive us into his favour; and finally, that he would establish the independence of these United States upon the basis of religion and virtue, and support and protect them in the enjoyment of peace, liberty and safety."

I do therefore by authority from the General Assembly issue this my proclamation, hereby appointing Thursday the 9th day of December next, a day of publick and solemn thanksgiving and prayer to Almighty God, earnestly recommending to all the good people of this commonwealth, to set apart the said day for those purposes, and to the several Ministers of religion to meet their respective societies thereon, to assist them in their prayers, edify them with their discourses, and generally to perform the sacred duties of their function, proper for the occasion.

Given under my hand and the seal of the commonwealth, at Williamsburg, this 11th day of November, in the year of our Lord, 1779, and in the fourth of the commonwealth.

THOMAS JEFFERSON[8]

The secular myth that the American Revolution was built on Enlightenment principles is debunked by the very words of Thomas Jefferson in his address: "that he would establish the independence of these United States upon the basis of religion and virtue, and support and protect them in the enjoyment of peace, liberty and safety."[9]

How has it come to this? It has come to this because Jews and Christian fell asleep at the wheel, and now they find themselves in the trunk of the car. It is not the fault of the progressivists, socialists, and collectivists. We should expect the enemy to always fight to gain control. It is our fault. It is the fruit of our apathy and the natural result of the horse blinds created by our materialism and hedonism and our subsequent acceptance of the postmodernist division of truth—the divided field of knowledge. America has been hoodwinked by the quite clever smokescreens that hide our true identity as a Judeo-Christian republic.

Francis Schaeffer visualized this division as a two-story house where there is an impassable floor between the two stories. The divided field of knowledge is the dualistic, postmodern, neoplatonic interpretation of reality whereby they categorize all objective science in what Schaeffer calls the lower story and all metaphysical sciences as the subjective upper story. Anything that has to do with morality or metaphysics is considered subjective and relativistic. It is the underpinning of their drive to remove all religious thought from law and replace it with a Darwinian materialistic religion.

The modern concept of the wall of separation between church and state touted as the rationale for removing any religious thought from government does not even come from the Constitution but from a misinterpretation of a personal letter written by Jefferson to the Danbury Baptist Church, which has been completely taken out of context.

At the time of the letter, on October 7, 1801, each of the states in our fledgling nation had accepted a particular denomination as its state religion. Imagine that! I suppose the states were also ignorantly unaware of the politically correct interpretation of the First Amendment.

The Danbury Baptist Church was at that time a minority religion in its home state of Connecticut. Church members were concerned that the state religion might in the future interfere with their church doctrines and cause them to veer away from their particular church ordinances.

The church was one of the strongest supporters of the Anti-Federalists who sought to keep the power of the federal government to its absolute minimum. I find it quite humorous that our modern atheists who are almost monolithically Super-Federalists love to prop up Jefferson as one of the few deists in the long list of our Founding Fathers who were devout Christians. If he were alive today, he would be leading another revolution against them as one of the strongest proponents of Anti-Federalism who championed state sovereignty and starkly warned us of the future danger of banking cartels.

The memory of the European problems in this regard was fresh in their minds. Hence, these members of the Danbury Baptist Church wrote to Jefferson, who was an avid Anti-Federalist, concerning their future fears in regard to their religious liberty under the Constitution of the State of Connecticut.

Our Sentiments are uniformly on the side of Religious Liberty—That Religion is at all times and places a Matter between God and Individuals—That no man aught to suffer in Name, person or effects on account of his religious Opinions—That the legitimate Power of civil Government extends no further than to punish the man *who works ill to his neighbour*: But Sir, our constitution of government is not specific. Our antient charter, together with the Laws made coincident therewith, were adopted as the Basis of our government, At the time of our revolution; and such had been our Laws & usages, & such still are; that religion is consider'd as the first object of Legislation; & therefore what religious privileges we enjoy (as a minor part of the State) we enjoy as favors granted, and not as inalienable rights; and these favors we receive at the expence of such degrading acknowledgements as are inconsistant with the rights of freemen. It is not to be wondered at therefore; if those, who seek after *power & gain* under the pretence *of government & Religion* should reproach their fellow men—should reproach their chief Magistrate, as an enemy of religion Law & good order because he will not, dares not assume the prerogative of Jehovah and make Laws to govern the Kingdom of Christ.[10]

Thomas Jefferson responded by assuring them that the First Amendment of the United States Constitution would not allow state governments to interfere with the religious rights of any individuals. His point was that no specific denomination can force itself upon the people through the government.

Believing with you that religion is a matter which lies solely between Man & his God, that he owes account to no other for his faith or his worship, that the legitimate powers of government reach actions only, & not opinions, I contemplate with sovereign reverence that act of the whole American people which declared that their legislature should "make no law respecting an establishment of religion, or prohibiting the free exercise thereof," thus building a wall of separation between Church & State. Adhering to this expression of the supreme will of the nation in behalf of the rights of conscience, I shall see with sincere satisfaction the progress of those sentiments which tend to restore to man all his natural rights, convinced he has no natural right in opposition to his social duties.[11]

His letter was designed to alleviate their concern that their state government could impose upon their church the religious doctrines of the particular state denomination. Jefferson responded by saying that the religious liberty of every individual is protected by the First Amendment, which not only keeps the government from forcing upon us a particular denomination but also protects our liberty to express and practice openly our religious beliefs in every aspect of our human endeavors, whether in our private or public life. The government is prohibited from taking away our right to freely express our religious views.

The idea that our laws would not be based on the Judeo-Christian worldview, which is, in fact, the very basis for the concept of individual rights, was never part of the thinking of any of our Founding Fathers. Anyone who wants to know the truth can read the transcripts of Congress that have been recorded since the beginning of our nation. Not a single word is ever mentioned in the Library of Congress that can be construed as remotely resembling the modern concept that divorces religious thought completely from law and the government. The very words of Jefferson in his Thanksgiving declaration debunk the secular myth that our government was not built on religious thought: "That he would in mercy look down upon us, pardon all our sins, and receive us into his favour; and finally, that he would establish the independence of these United States upon the basis of religion and virtue, and support and protect them in the enjoyment of peace, liberty and safety."[12]

Today, we are buffeted by two extremes that threaten to take our religious liberty, which is enshrined as inviolable by the First Amendment. At one end of the spectrum, we find that naturalists adamantly reject any association whatsoever between politics and religion and thus force upon us their own atheistic religion through government and our educational institutions.

At the other end, we find Muslims who teach that the political system should be ruled by the religious institution as a theocracy.

Both extreme views are wrong. Both these views lead to tyranny. Both these views rob a sector of society from rightfully expressing their God-given religious rights. Both these views are anti-Judeo-Christian in every sense of the word.

However, the idea that church and state could be absolutely separate is simply an illusion created by the clever semantics of progressivists. These are the smokescreens that cloud the truth and deceive the uninformed. These are the clever semantics used by the Enemy of Man to disarm the church and keep us hemmed in and silent so he can achieve his nefarious globalist agenda.

The reader must understand that the idea of separating our religion from our politics is simply an unabashed oxymoron. Every person on this planet adheres to a religion, even when that worldview is anti-religious, such as atheism. It is nonetheless a form of religion, for it establishes a definitive worldview from which a set of ethics springs forth.

Hence, the foundational worldview of any individual is, in essence, his or her religion, and it invariably impacts his or her political decisions. This irrefutable premise is even supported by the 1961 *Torcaso v. Watkins* U.S. Supreme Court decision that declared humanism to be a religion equivalent to any theistic or monotheistic religion. Thus, at least in America, the highest court has correctly stated with clear certainty that our presuppositions, whether theistic or atheistic, are in fact, our religious creed.

All foundational worldviews will in every case impact our political direction. And the objective observer must come to the only conclusion that there is no way to divorce the two. It is impossible to formulate any law without the influence of a foundational moral or amoral principle behind it.

As its stands today, only the relativist, the atheist, is free to express his or her religious view in government since God has been effectively kicked out of our laws. Their anti-God religion is the only religion that can be taught in our public schools. All attempts to teach an alternative to the evolutionary atheistic Darwinian model are essentially prohibited by this clever smokescreen. America has been hoodwinked. This must end. Scientists who believe that the design of our universe and especially life requires a designer must be allowed their God-given freedom of expression in academia.

The freedom of the expression of all religions should be assured and protected at all costs in any form of government devised by people and especially so if this form of government is to reflect Judeo-Christian principles. God does not call us to force religion on others but to give all humanity the freedom to choose their religion individually. In the end, they will answer to God, not humans.

But in the area of morals and the establishment of a legal system that can assure equal justice to all, there can be no equivocation that our nation was founded specifically to reflect the Judeo-Christian worldview in this regard. These were from the very beginning predicated on the absolutes of God's truth and not on a relativistic plurality.

For this reason, our Founding Fathers did not establish a pure democracy but rather a constitutional republic. America did not establish a form of government in which the law was subject to popular trends, but to an absolute standard of justice that cannot ignore the individual rights of the minority. That is critical to understand.

Our form of government was not built on the foundation of the Greek city-states, as you have been taught in secular public schools. In a pure democracy, the freedom of the individual is suppressed and subjugated to the needs of the collective or majority. There are no rights for the minority in a pure democracy. There is only the will of the majority.

For this reason, our Founding Fathers also instituted the electoral college for presidential elections in order to ensure equal representation of all states, regardless of their population. Without the electoral college, the presidential election would be controlled by eight states, including California and New York, which have the largest populations and would infringe upon the rights of the smaller individual states. Today, this vital provision is under attack by the socialist movement in the Democratic Party.

Sadly, our nation is being led away from that vision for which so many have died to create and preserve. It is for this reason that the United States was formed as a Judeo-Christian republic whose absolute standards of law were gleaned from the bedrock of the Holy Scriptures.

It was not established as a true democracy, which is the reflection of the majority rule, *consensus gentium*. But instead, as a representative republic, it maintains both the freedom of the individual and secures justice for all, predicated upon an absolute standard of morality and the inherent rights of all individuals secured by our foundational Constitution. Only as a direct result of this worldview can minorities as well as all individuals and states, be protected from having their rights infringed upon by either the majority or the power elite.

Our government was founded on biblical principles that ensure the rights of all citizens as equal and subsequently provide protection for the minority. Much of our jurisprudence was modeled after *Commentaries on the Laws of England by Sir William Blackstone (1723–1780)*:

> By the time the Declaration of Independence was signed, there were probably more copies of his *Commentaries* in America than in Britain. His *Commentaries* shaped the perspective of American law at that time, and when you read them it was exactly clear upon what that law was based.[13]

Our nation was built on a constitution whose legal principles were garnished from the Word of God as the absolute standard for truth and justice. Without the foundational basis of the absolutes of scripture, there can be no feasible way of adopting any legal standard that would serve to ensure fairness and freedom to all the individual citizens in any form of government.

Without an absolute basis for moral laws, justice is replaced by tyranny as power becomes right and the popular whims override the rights of the minority. Thus, the naturalist who seeks to drive an impenetrable wedge between religion and politics is simply advocating for his or her particular brand of anti-religion to be preeminently represented in the legal system at the exclusion of all others. Progressivism-socialism is pushing for a fundamental shift in our jurisprudence that will eventually destroy the rights of the individual for the sake of the collective.

Socialism is being rammed down our throats by the Hegelian relativists who greedily push to hoard and centralize political power and exclude all others. Such a unilateral force will inevitably erase the freedom of the individual, which is paramount to the Judeo-Christian worldview and the very foundation of our American constitutional republic. It will destroy our free market system and make all citizens financially dependent on the federal government. Such a policy would ensure that only the atheistic worldview is represented by our government. In spite of all their flowery propaganda and deceptive rhetoric about being multicultural and pluralistic, it is the exact antithesis of true pluralism.

But we must be careful to make laws only on the basis of the absolute moral standard of God and not on our own cultural biases on issues that are not absolutely stipulated in scripture. The function of government is to create laws that are predicated on the absolute bedrock of the moral mandates of the creator in order to ensure equal justice and the freedom of all its constituents.

It is not to force one particular brand of religion or denomination on humankind but to provide a safe haven for the free practice of all religions. These religions can be permitted as long as they do not interfere with the freedom of others and as

long as they do not transgress the foundational laws stipulated by the U.S. Constitution. That is the basis for a true constitutional republic.

Here lies the important difference between the Judeo-Christian concept of justice and the concept derived from Islam, which seeks to force its religion on others by brute force, completely countermanding God's intent to give humankind a free will, as Jefferson clearly stipulated in his response to the Danbury Church. Faith is not a matter of birthright, nor is it exclusively a matter of the heart. Faith is a matter of the heart and reason. If there is a God, then the universe has order and congruence, and the minds of people cannot be sidelined as superfluous.

Reason cannot be ignored. True faith is rational and is in congruence with true truth and the reality of the space-time continuum we inhabit. Moreover, if God created humans as beings and not as unthinking automatons, then the ability to choose through rational processes is an intrinsic characteristic of humans as they were designed by God.

Any religion that forcefully imposes its dogma, circumventing the intrinsic right of all human beings to choose of their own volition and as their own consciences dictate, is contrary to the design of the creator and an illegitimate mandate that contradicts God's design for humanity and government.

Contrary to the Muslim position, it is the will of God that humankind have a free will to enable them to come to Him by faith and not by the point of the sword. He does not need anyone's feeble hand to bring light to those who would seek Him.

The false dichotomy created in humanity between Dar al Islam (the House of Submission) and Dar al Harb (the House of War—infidels) by the Islamic religion that revokes all human worth and intrinsic value to the person they label as an infidel is in direct conflict with the will of the one true God and creator of the heavens and earth. He created all human beings in His image to have infinite worth and value, regardless of their faith, intelligence, ethnic roots, economic stature, or any other human superficial designation.

Our human dignity comes not from our intelligence, possessions, religious views, or any other choice we make. It comes from the breath (*nashama*) of God as He created human beings in His image. This is the rock-bottom fundamental premise instituted by God for our entire human civilization, and it cannot be compromised without creating the most heinous direct affront to God and to the sanctity of human life. The sanctity of all human life is an irrevocable and universal axiom established by the creator. Any nation or religion that transgresses this fundamental axiom is in direct conflict with the creator.

God calls us to reason with Him and come to the knowledge of His true truth by the individual act of faith and not at the point of the sword. Islam attempts to force upon all others its cultural relativistic norms devised by its clerics with equal adamancy to what they claim to be God's direct revelation. We must never elevate our human dictates to the same stature as God's revelations. This is also a direct affront to God.

The Pharisees and the scribes asked Him, "Why do Your disciples not walk according to the tradition of the elders, but eat their bread with impure hands?" And He said to them, "Rightly did Isaiah prophesy of you hypocrites, as it is written:

'THIS PEOPLE HONORS ME WITH THEIR LIPS,
BUT THEIR HEART IS FAR AWAY FROM ME.
BUT IN VAIN DO THEY WORSHIP ME,
TEACHING AS DOCTRINES THE PRECEPTS OF MEN.'

Neglecting the commandment of God, you hold to the tradition of men."

—Mark 7: 5–8

We must not allow Islamic radicals to warp this ideal with their erroneous and legalistic views that blur the distinctions between moral absolutes and personal cultural preferences. Their view of the intrinsic rights of women is abysmal and barbaric. Many Islamic countries do not even allow women to get an education and consider them nothing more than the personal property of men. This invariably results in a tyrannical government that strips sectors of the society, as well as individuals, from their God-given rights to life, liberty, and the pursuit of happiness.

Justice is not served by allowing only atheists and agnostics to impose their brand of immorality on society, which also robs sectors of our human family such as unborn human beings of their individual rights. If those who hold to the Judeo-Christian worldview do not exercise their right and duty to vote and influence our government in the direction in which we believe it should go, then we are abrogating our responsibility to God in becoming the salt of the earth and in rendering unto Caesar that which belongs to Caesar. In other words, God wants His children to participate in government.

If we do not, we are not only failing in our civic responsibility to be a participating citizen of our nation, but we are also failing in our sacred moral and religious duty to be the defenders of the downtrodden and abused. To abstain from the fray is to abrogate our civic duty as law-abiding citizens of our country, which morally constrains us to vote our conscience. In essence, we are then abdicating control of the direction of our government to those who champion the naturalistic worldview.

The Neo-Platonic Artificial Division – The Separation Deception

The major problem at the root of the abortion dilemma and all the other social issues of our Western culture that will eventually directly result in the loss of our freedoms in America, stems from the fact that those who hold the Judeo-Christian worldview have retreated from competing in the public square. We have been duped into accepting this neo-Platonic, politically correct ideology that separates the religious from the political sphere. The natural outcome of this division is the false assumption that we should not legislate morality. We have already established that contrary to this deceitfully crafted smokescreen, all legislation is, in fact, a reflection of a moral stance. This neo-Platonic, artificial division of reality developed in our society as a direct result of an occult doctrine bandied about for thousands of years.

The deterioration of these intrinsic human rights and freedoms in our world today is the direct result of the influence of the Gnostic, neo-Platonic heresy that has quickly risen to overshadow our Judeo-Christian heritage and now dominates our educational institutions and political system. Those who hypocritically claim to champion tolerance are the most rabidly intolerant of all.

Plato incorrectly created a false division in reality when he described the spiritual realm and the physical realm as two separate and disconnected entities. That is incorrect. The spiritual and the physical are both parts of the same reality and under the lordship of God.

The naturalists, who want to force upon us their own brand of religion, have conveniently championed this false dichotomy that seeks to separate any religious ethics from government. Those who espouse a Hegelian worldview have gained the upper hand in political offices and academia and have caused our nation to digress radically from the original precepts and hopes of our Founding Fathers.

The absolute precepts of justice, which alone can create freedom and liberty, have been slowly replaced by relativistic mores and situational ethics, which can only lead to tyranny and the inevitable abuse of the masses. We must walk circumspectly if we are to preserve these vital freedoms.

The reader must understand that every single law passed by our legislatures is reflective of some moral framework. There is no real division between the philosophical presupposition of the legislators and the laws they impose. It is out of their philosophical worldview that people develop notions of justice and morality. Therefore, they cannot be segmented into this artificial separation created by those who want to silence Jews and Christians.

If Jews and Christians do not advocate for laws that reflect the moral standards of the Holy Scriptures, then the laws of our country will simply reflect the relativistic standards of the naturalists, pantheists, and occultists who fervently advocate their particular notions of relativistic justice. Many Jews and Christians have equally failed to understand our responsibility to do our part as citizens of our nation so justice and freedom for our fellow citizens are ensured.

Moreover, our particular form of government has entrusted every American citizen with the perpetual duty to protect and maintain these foundational precepts that were imputed to our republic at a great price by our Founding Fathers. To forego this responsibility to our government runs against the very principles of both Christianity and Judaism and makes us irresponsible and negligent citizens within our community, regardless of our religion.

In effect, Christians and Jews have been sidelined from the public square through the use of this clever smokescreen. We have failed to understand that the only real basis for human value and the rights and responsibilities it entails stems from the singular fact that we have been created in the image of God. We have failed to understand our true identity. This is not just an American problem. It is a global problem that threatens all humanity in every nation.

We must understand that the noble aspiration of creating a government that assures justice and freedom for every member of society without regard to rank or social status or the color of their skin or their religious creed was predicated on the worldview espoused solely by biblical absolutes that uphold the unique and infinite value of all humans because they have been created in the image of God. Were it not for this indispensable foundation, these absolute principles would not have been part of our U.S. Constitution.

It may be argued that the execution of this noble aspiration was imperfectly carried out, but none can deny the value and revolutionary uniqueness of this experiment to create a form of government in which the rights of the individual are secured while the power of the government is derived from the governed. Without these two primordial elements of our original foundational worldview, government would be reduced to the rule of the mob or the tyranny of a despot or oligarchy. This is our religio-political legacy in America. Furthermore, it is our responsibility to continue to maintain these freedoms through our concerted and circumspect civil and political activity.

Jews and Christians must not fail to engage in the political process in order to influence our governments toward the principles expounded by the Judeo-Christian worldview, which alone can lead to justice in our nation and on our planet. Modern man has erroneously compartmentalized our Judeo-Christian experience from our daily practical life. The spiritual realm has been separated from the physical realm. Thus, the Judeo-Christian faith and our social sphere, which includes politics, science, and our educational system, have been artificially segregated.

Modern man, including Christians and Jews, has in this regard completely missed the boat. The physical and the spiritual are not two separate counterparts of reality. They are a single continuity of the universe created by God. That is not only metaphysically true but also empirically and physically true.

This continuity is evidenced by reason and not by blind faith. We can cite the modern string theory as mathematical evidence that the invisible and the visible are parts of the same reality. The string theory, or to be more specific, the M-theory, specifically stipulates that there are three physical dimensions that are visible, while there are seven physical dimensions that are invisible. All 10 of these dimensions are found within the dimension of time. Hence, there is continuity to the diversity found in our universe that is reflective of the creator's design, which ties the particulars to the universal.

Naturalists often incorrectly insinuate that these dimensions are separate universes. They are not. They are interconnected, and the mathematical equation that describes them is also the empirical evidence of their interconnection and unity. There is a major difference between a dimension of our universe and a separate universe, which by definition could have no interconnection with ours.

God is sovereign over our entire life, and His absolute precepts are the matrix of all reality. To shut Him out of any area of our lives is to shun His sovereignty by making humans gods over that specific area. This Platonic heresy, which artificially separates reality into two separate spheres—the spiritual and the mundane—has been promoted by the occult. But in God's economy, no such thing exists.

All reality is a continuum under His lordship. The seen and the unseen are both integral parts of the same reality. Our ability to perceive them does not provide the proper rationale to divide them as separate and distinct realities any more than we can claim there is no connection between the seven invisible spatial dimensions and our three visible spatial dimensions. Furthermore, our responsibilities as believers and citizens in both of these areas ought never to be in contradiction. That is what religious freedom means.

The Kabbalists and Gnostics have bandied about this dualistic occult view since ancient times, and it continues to this present day in the fashionable neo-Platonic ideology that has been accepted by our modern Western culture. The physical realm has been associated with reason, while the spiritual realm has been labeled as mythical, outside of reason, and therefore relativistic.

The realm of faith and religion has been consequently relegated to a purely subjective and inferior compartment, while we have been deceived into thinking that we can find the answers to all of humankind's woes strictly through the physical realm. Science has become our new god. And it is not true science but rather a particular brand of science that promotes the religion of naturalism, which can accurately be labeled the cult of scientism.

Our Founding Fathers had a vision that was restated by Jefferson in his letter to the Danbury Baptist Association, that our nation would always foster the "progress of those sentiments which tend to restore to man all his natural rights."[14] If we do not involve ourselves in our civil duties and public responsibilities, then how can we hope to secure this great American dream? That vision was clearly established in the Declaration of Independence:

> We, therefore, the Representatives of the United States of America, in General Congress, Assembled, *appealing to the Supreme Judge of the world for the rectitude of our intentions*, do, in the Name, and by Authority of the good People of these Colonies, solemnly publish and declare, That these United Colonies are, and of Right ought to be Free and Independent States; that they are Absolved from all Allegiance to the British Crown, and that all political connection between them and the State of Great Britain, is and ought to be totally dissolved; and that as Free and Independent States, they have full Power to levy War, conclude Peace, contract Alliances, establish Commerce, and to do all other Acts and Things which Independent States may of right do. *And for the support of this Declaration, with a firm reliance on the protection of divine Providence,* we mutually pledge to each other our Lives, our Fortunes and our sacred Honor (emphasis added).[15]

There was no separation between God and our political system when our nation was forged. The Founding Fathers expressly appealed "to the Supreme Judge of the world for the rectitude of our intentions." In other words, it was God they considered to be the supreme measure of righteousness and justice, and to Him alone they looked for approval of this new form of government. But it did not end there. They also depended on Him and Him alone to protect this form of government in the future, "with a firm reliance on the protection of divine Providence."

Any subsequent governmental institutions that are legally subservient to the preeminent document that serves as America's articles of incorporation cannot abrogate those deep sentiments and foundational principles. Any legislation or judicial mandate that transgresses the primordial document that legalized our nation is therefore automatically illegitimate and null and void.

However, we must walk circumspectly here and not fall off the other precipice by becoming manipulated by the self-serving purviews of any single political party. God is neither a Republican nor a Democrat nor an American, for that matter. We must not allow any human institution to manipulate us for its particular vested interests. This has been the unfortunate rule and not the exception of our human experience throughout history. And equally so, we must not wrap our American flag around God. God is lord of all reality and every nation, people, and tongue. He is not our exclusive property.

Whenever a given party stands on issues that represent our Judeo-Christian worldview, we must support and acknowledge it. But when it opposes those principles, we must oppose that party. We must acknowledge and insist that any legitimate, human form of government cannot compromise any of our basic human rights as ordained by the creator.

While we are charged with the responsibility to submit to the authorities, but that does not give government carte blanche to move beyond the limits prescribed by God or by the Constitution, which provides the limits of their power. Our human governmental institutions ought to be limited. Their powers should be restricted. Their funding should be constrained. Their authority should be restrained. Their dominion should be curtailed. Their offices should be checked and balanced so they are impeded from transgressing our basic individual rights given to us by the creator. The politicians are accountable to "we the people" who consent to give them power to serve us and not the other way around. They are not our masters; they are our servants.

Here is the great danger of the centralization of power of the federal government that has been illegitimately accumulating in size and power over our individual lives without any public outcry to oppose it. It is the danger of all dangers to our freedom and individual rights. We must resist the burgeoning growth of the federal government. We must become aware of its clever subterfuge. We must understand the clever tactics and smokescreens that it uses to consolidate its power in the guise of helping the common citizen.

We must demand that our government balance the budget just as every one of its citizens must do. As of March 2019, the nation's national debt has ballooned to a whopping $22 trillion, and the interest on that debt has increased 400%. International bankers are laughing all the way to the bank. It is a critical juncture, and our nation stands on the brink of a fiscal cliff. If we fail to address that, our economy will eventually collapse, bringing disaster to our children. Such a collapse, of course, would be the perfect ruse to establish a global monetary system, which the globalists are hard at work to obtain.

Do not fall for the socialist siren song of "free stuff" that will demand higher taxes and destroy the middle class, for it is also intended to bring our nation to fiscal collapse by exploding our national debt so they can institute their globalist agenda. Our sovereignty as a nation is at stake and with it our Bill of Rights and our ideal of self-government.

We must not fall prey to this clever smokescreen. We must not allow the progressive movement, entrenched in Washington, DC, and academia, to rob us of our precious freedoms given to us through the wisdom of our Founding Fathers. We must not allow them to rewrite our Constitution through the grid of their diabolical collectivist ideology.

The Great Evangelical Disaster

The real problem does not come from the fact that the naturalists are promoting their worldview. We must expect this in a fallen world. The problem is that Christians and Jews have retreated from the public square. They have abandoned their responsibility to stand fast against the cosmic forces that would bring tyranny to our human family.

The sad fact is that Satan has lulled both the churches and the synagogues into the comfort zone where the only thing that is of real importance to most in mainline Christianity and Judaism is the goal of maintaining personal peace at any price. We have been deceived. We have been chasing the "mirage," mesmerized by the false promise of the Great Deceiver.

Sadly, even many Christian pastors and Jewish rabbis are more concerned about building their own kingdoms than the kingdom of God. Few are the pastors and rabbis who are willing to speak on social issues from the pulpit for fear that they might offend someone and impact their collection plate or their precious building projects to enshrine their own little kingdoms. This was certainly the case in Nazi Germany. It is nothing new.

Of course, they rationalize this myopic and selfish decision with a variety of excuses, but in the end, most of them are colored by the neo-Platonic heresy. Some justify this false dichotomy, citing the New Testament: "Render therefore unto Caesar the things that are Caesar's; and unto God the things that are God's" (Matt. 22:21 KJV). They fail to realize that they are not rendering unto Caesar when they do not participate in the public square. They fail to realize that God is the one who raises up kings and takes them down.

Moreover, God is lord of every area of our lives, and when they divorce our spiritual responsibility from our political responsibility, they are denying this basic reality and acceding to the neo-Platonic worldview, which creates this false dichotomy. Francis

Schaeffer was clear about this in his prophetic warning to us in the last century:

> Accommodation, accommodation. How the mindset of accommodation grows and expands. The last sixty years have given birth to a moral disaster, and what have we done? Sadly, we must say that the evangelical world has been part of the disaster. More than this, the evangelical response itself has been a disaster. Where is the clear voice speaking to the crucial issues of the day with distinctively biblical, Christian answers? With tears we must say it is not there and that a large segment of the evangelical world has become seduced by the world spirit of this present age. And more than this, we can expect the future to be a further disaster if the evangelical world does not take a stand for biblical truth and morality in the full spectrum of life. *For the evangelical accommodation to the world of our age represents the removal of the last barrier against the breakdown of our culture.* And with the final removal of this barrier will come social chaos and the rise of authoritarianism in some form to restore social order.[16]

His prophetic warning is becoming a stark and foreboding reality. The collective accommodation, which dominates our culture, is the result of individual accommodation. The churches and synagogues are sick because our leaders are sick. We have lost our true identity as the bearers of the torch of truth, justice, holiness, humility, and righteousness.

Our over-preoccupation with materialism and recreation, the pursuit of pleasure, or the pursuit of success to fan our egos has increased to the level of being almost an obsession within our modern culture. We have been neutralized by the pursuit of the mirage.

Sadly, the increasing decadence of our Western culture has given Islam ample fuel to attack us. In this they are absolutely correct, and our apathy will be our eventual undoing. We have no time in our lives for anything that could bring on us the necessity of some form of sacrifice. We shun the prospect of any endeavor that would create even the slightest conflict or discomfort. And we must acknowledge that we flee from this responsibility as if it were the Plague.

This, my friends, is nothing less than cowardice. Perhaps we think we are more sophisticated than those in the past who worshipped wooden and stone idols. But in reality, we are no different. Our idols may be different, but we are an idolatrous and perverse generation nonetheless.

The battle begins internally as each of us understands the futility of chasing the ever-elusive mirage. What is this mirage? It is the idea that we can heal ourselves without God. It is the false notion that we can be self-sufficient. It is the idea that we can find happiness in the accumulation of material goods or the accomplishments or the rewards garnered in the physical realm. And this applies not only to the individual but also to our ministries and our nations at large.

We have become engrossed in our own kingdom-building projects and have lost sight of our true calling from God to build His kingdom. Our individual and collective wholeness comes only by consistently applying God's standards to our individual lives, families, jobs, and communities.

Only then will we be freed from the enslavement of the magnificent lie—the lie injected into our space-time existence in the Garden at Eden, the lie that deceives us to think we can be gods. We are not gods, but we are also not a worthless zero, as Schaeffer points out. We have infinite worth and value since we are all created in the image of God.

Our intrinsic individual value does not come from our material possessions, our physical appearance, our abilities, our accomplishments, or our positions of power. This is simply chasing after the mirage. We are valuable because of the *Imago Dei*—the image of God—in us and because He loves His children unconditionally. Therefore, we have a basis from which we can learn to love ourselves and others in a balanced form.

True self-love stems from the love of God, which then frees us to love others. Whoever hates their brother does not know God, for God is love. Those who, in the name of God, hate, kill, and torture do not worship the true creator of the universe but rather a sordid and malevolent counterfeit created in the minds of demons and selfish men.

> ***The one who does not love does not know God, for God is love.***
>
> ***We love, because He first loved us. If someone says, "I love God," and hates his brother, he is a liar;*** *for the one who does not love his brother whom he has seen, cannot love God whom he has not seen. And this commandment we have from Him, that the one who loves God should love his brother also* (emphasis added).
>
> —1 John 4:8, 19–21

If we say we are the people of God, then we cannot do anything less than love our brothers and sisters regardless of their ethnicity, education, financial status, gender, age, or any other superficial distinction we can artificially make. That does not mean, however, that we must be tolerant of wicked ideologies that enslave our brothers and sisters. We are to hate sin but love the sinner.

Love means selflessness. Whoever cares only to appease their own desires will never be fulfilled, no matter how successful they become at their quest, no matter how much of the mirage they chase. No matter how far they have traveled on that trail to the mirage, it is never enough. They will remain shackled to the lie and to the Father of Lies.

Do not misunderstand me. God is not against comfort or recreation. After all, He is the one who instituted the Sabbath Day of rest, a principle that is little regarded in our modern culture. We are commanded to leave this day holy, to appoint every seventh day to the worship of the creator and not to the accumulation of wealth. But when our selfishness rises to the point that we neglect our responsibility to God and to our brothers for the sake of comfort and recreation, it becomes an illegitimate pursuit that in the end will not bring us either peace or lasting comfort. It has become idolatry!

Taking a stand for truth more often than not requires some sacrifice. And it is that realization that paralyzes the selfish from getting involved in the eternal cosmic battle between the forces of good and evil. God calls us to be separate but not isolated. God has not called us to be the sugar of the world; He has called us to be the salt. It may not be very pleasant to put salt on a wound because it stings. But it disinfects and preserves.

I surely detest the prospect of any form of conflict. But I have learned that in spite of my wishes, if I do not address evil promptly, the conflict will eventually come to me in a much greater form. If I shun my responsibility to oppose the encroachment of evil from the onset, it will not disappear. It will only grow larger and stronger and more difficult to overcome. And in the end, my cowardice will cost me more than I could have imagined.

As children of God, we are not to place our trust in any political or even spiritual leader, but solely on God. Those well-meaning but deceived pastors who teach that our only spiritual responsibility is to reach the lost are, in fact, aiding the enemy of humankind by infringing on our civic rights and duties to influence our government toward the righteous and just principles of our Judeo-Christian worldview. They will be held accountable to God. They may think they are being spiritual, but they are not. One responsibility does not exclude the other. Both are required. We must render unto Caesar that which is Caesar's and unto God that which is God's.

They are being deluded into aiding the wicked to gain power and promote injustice. They are promoting the dualistic deception created by the Great Deceiver and emasculating believers from the God-given right and privilege to participate in the task and duty of citizens to give their consent to our rulers as proscribed by the Declaration of Independence and the Constitution. They are, in fact, contradicting the Apostle Peter's command to be obedient to the governmental authority that God in His providential grace has afforded us in this great nation.

The Battle of the Long Defeat

Make no mistake; we are at war. It may not be a war fought with guns and tanks, but its outcome will be of the gravest importance to humanity. And neglecting it will be utterly catastrophic for our children and our children's children. It is a war for the minds of human beings. The defeat of the Enemy of Man may not come in our generation, but the battle will surely conclude with its defeat at the consummation of the ages. The victory has already been won.

The ancient cosmic battle between the forces of evil and good is raging on our planet with mounting force as we near the arrival of the long prophesied global ruler—the Antichrist. The battle in our generation rages on two fronts: the Middle East and the West. From the Middle East, radical Islam seeks to force upon all humanity the worship of their demon god Allah and the forced institution of their barbaric Sharia law. From the West, we have the progressivist-socialist-collectivist ideology that seeks to dissolve all nations into a global government, which would be quite lucrative for the elite financial merchants who are the real power behind this movement.

The promoters of this global government have been carefully preparing the philosophical foundation from which their political ideology can be achieved. Of paramount importance to their global aim is the task of annihilating or neutralizing any political structures based on the foundation of a legal system that reflects absolute standards.

Their goal is for governments that are founded on the absolute principles espoused by the moral law in the Holy Scriptures to be changed to a relativistic system in order to bring all nations under a unified global governmental structure that can then be effectively manipulated by the elite in power. Therefore, it is of paramount importance to these globalists to fabricate a wedge of separation between the government and the Judeo-Christian worldview. This is the separation deception that has been disseminated throughout our nation through public education and the mass media.

To this end and at the opposite extreme, Satan has fostered the Islamic, corrupted form of God's true religion in order to foment hatred toward an absolute moral law. Islam is the very tool of Satan to drive the world toward a naturalist worldview and a globalist police state that will promise to provide freedom from terrorism for all humankind. The Enemy of Man does not hesitate to use both ends against the middle.

Know this: If our nation is able to resist the socialization process that has engulfed the nations of the world, then America and Israel will be hated above all nations. We will become the social outcasts of this new world order. Even now, the seeds of this fulminating hatred are being sown throughout the entire world.

We have three choices. We can choose to remain true to the Judeo-Christian principles and become hated by all nations. We can give in to the relativists and become absorbed by the nations and lose all our cherished liberties afforded to us by our Constitution. Or we can choose to ignore the radical Islamic colonialist aggression that is mutating into a global army that will destabilize all nations and bring in their savage religion by the force of their sword. There is no fourth choice. That is the political reality that faces our generation. If we withdraw to our churches as monastic orders of isolationism and step out of the public square, we will have no one to blame but ourselves, and our children will one day hold us accountable for not securing their future freedoms.

We can see this subliminal animosity toward America forcefully mounting throughout Europe, South America, and Asia. Why? Because our nation has been established as a form of government that builds its laws upon the firm foundation of the Judeo-Christian scriptures. We have historically stood for absolute standards in a world that is rapidly shifting toward an all-inclusive and syncretic faith built on the ever-shifting sands of relativism. To this end, the enemy of God will always seek to sabotage our economy to reduce and hinder our power to resist this globalist aspiration that his demonic hierarchy is promoting in all nations.

If we fail to resist the globalist aspirations of both these erroneous views from the Middle East and the West, we will have no one to blame but ourselves. Israel is further down this slippery slope than America; secular forces have entrenched themselves in Israel's government. But trust me, the world will not accept Israel. Satan hates Israel. Through Israel came the prophets and the Messiah. The world will seek to sacrifice Israel for its diabolical global government. But Israel belongs to God, who will have something to say and do about that.

The freedom and prosperity that we have enjoyed in our Western culture is the product of the Judeo-Christian foundation that established it. To be sure, it has been imperfectly executed, and we have strayed far from the original foundation. But when it has gone wrong, it has been because it has strayed from its foundation, not because the foundation was wrong.

Today, we face a double globalist threat from the West and from the Middle East. We face the globalist agenda engineered by the naturalists and the occult, which have risen to power throughout the Western nations. And we also face the globalist agenda of the radical Islamo-Fascists who are also rising to power through the enormous profits of the oil industry.

We must not flinch at our responsibility to intelligently and lovingly confront these two erroneous worldviews that will otherwise inevitably embroil our planet in a bloodbath. We must not allow the neo-Platonic false division of reality to disable us in the political process. We must not deceive ourselves into thinking that God is not necessary in human government. This is nothing more than a reflection of the error introduced into our space-time continuum in the Garden of Eden. The clever Dragon deceived our mother Eve and convinced her that she did not need God. He convinced her that knowledge could make her equal to God. Moreover, his original motive for introducing this magnificent lie has never wavered. It is his goal to usurp the throne of God both in heaven and on earth until the day he is thrown into the lake of fire.

The Magnificent Lie

The magnificent lie is the idea that knowledge in the horizontal plane is all we need. It is the idea that self-interest is the highest goal of human endeavors. It is the lie that tells us we can achieve godhood through special knowledge that empowers self. This esoteric knowledge is what brings the followers of Lucifer into power. This has been the weapon of the Enemy of Man to ascend to the throne in heaven. He offers this esoteric knowledge and earthly power to those who do his bidding, sometimes without them even realizing it. The scriptures warn us that he seeks to ascend to the heavenly throne. But he also seeks to ascend to the global throne on our planet.

***

Beware! Satan never offers you anything without taking 10 times more away. He does not mind giving you what you think you want and then taking away what you really need.

***

The scriptures forewarn us that the Great Deceiver will smite the people of faith in wrath with a continual stroke. The watchers, who are under his control, will in the end of days rule the nations in anger, knowing that their time is short.

We must walk circumspectly, my friends, for the time of Jacob's trouble is rapidly approaching. Anger will ascend to the global throne when smoke rises like a furnace from Sheol. We need but to open our eyes to see that anger, which is already enveloping our planet under its dark, wicked cloak. Islam is prima facie evidence of this mounting supernatural rage on our planet. The wickedness of its brutality is so horrendous that it is surely a supernaturally fomented rage.

But our responsibility remains the same regardless. We are to stand for truth no matter the consequences or the probability of victory in our own generation. Our duty is to be faithful and not act solely predicated upon the calculations of our chances of success. Nevertheless, we must never lose sight of the fact that the Shade Slayer will come to break the coming night by the piercing light of the Morning Star rising from the East. Darkness cannot conquer light. The final victory is already ours.

It is written that the scepter of the wicked shall be broken. How will the oppressor cease? By the roar of the Lion of Judah, and in that day, the golden city of Babylon shall be no more. The roar of the Lion of Judah shall be wrath to the enemies of God, but it will not be wrath to the believers. The Lion of Judah will come to dwell among us:

"For I am God, and not a man—
the Holy One among you.
I will not come against their cities.
They will follow the LORD;
he will roar like a lion.
When he roars,
his children will come trembling from the west.
They will come from Egypt,
trembling like sparrows,

from Assyria, fluttering like doves.
I will settle them in their homes,"
declares the LORD.

—Hosea 11:9–11 NIV

Jerusalem will become His seat of power in the third Temple. The City of Truth will survive to rule over all the earth with the Son of David on His throne. In that day, the Great Impostor will be taken from the global throne and punished for his crimes against God's chosen people. All of the chief ones of the earth will rise up from their thrones and say to him, "Art thou also become weak as we?" His pomp will be brought down to Sheol, for he will be cut down without human agency.

How art thou fallen from heaven, O Lucifer, son of the morning! how art thou cut down to the ground, which didst weaken the nations! For thou hast said in thine heart, I will ascend into heaven, I will exalt my throne above the stars of God. . . . I will be like the most High.

—Isa. 14:12–14 KJV

Do not fail to see the deception of this megalomaniac mirage that infects us with false visions of grandeur, arrogance, bluster, and self-aggrandizement. Those who believe they have no need of forgiveness are those who deny the very Word of God and the judgment that entered into the Garden of Eden and brought death to humankind. Do not fail to see the deception of the magnificent lie that tells us we can be gods. If we fail to see the connection between the spiritual and the physical, we will become the victims of the ancient cosmic battle that has raged from the very beginning of our universe.

But we are not as those who have no hope. We are not children of the night. We know that in the end, true truth will be victorious. Ours is the privilege to join in this cosmic battle to illuminate the darkness with the light of truth, which alone sets people free. This we must faithfully do in every generation until the day our Champion comes to claim His throne.

The moral precepts of true religion form the only real foundation for a system of government that can ensure true justice and equality to its people. Here is where the battle rages in our generation. But the church has formed a moat around its castle and raised the drawbridge. Christians are content to sing hymns to one another while the world burns around them. God will hold us responsible as the unfaithful servant who buried his coin.

I must again insist on making clear that I am not advocating a theocracy. I am advocating obedience to our God. It is true that in our present epoch, church and state should be separate in structure, but church and state should not be separate in law, citizen participation, and philosophical foundation.

As John Whitehead has so brilliantly championed, the idea that law and church can be separated is but a carefully manufactured illusion created by those who wish to neutralize the influence exerted on our form of government by those holding a Judeo-Christian worldview.

But the time will come when the absolutes of God will be the rule of the entire planet. The scriptures have long prophesied of a day when peace will reign supreme and justice will rule our planet. That time is when the Messiah, Jehovah Tsidkenu, returns with the iron scepter of Judah to sit upon the throne of David. He will rule with righteousness over the entire world from His throne in Jerusalem.

His form of government will be the first to have true global justice. It will not be a democracy that is ruled by popular opinion nor will it be a tyrannical system predicated on maintaining the elite in power and exploiting the masses in order to do so. It will not even be a republic. It will be a theocracy based on a loving, righteous, just, and benevolent ruler. It is the only way a theocracy can work. No humanly contrived theocracy can work because no human is God, and all of us are susceptible to corruption.

Truth, righteousness, justice, and peace will reign supreme because the Messiah will judge justly. And every human on earth, regardless of his or her appearance, intelligence, color, or nationality will be equally treated and loved. For the first time in our history, all of humanity will truly have equal opportunity and true justice.

There will be no more armies and no more wars. Peace will rule the planet as Shiloh rules from Jerusalem. In that day, at long last, the City of David will be known as the City of Peace and the City of Truth.

Such a theocracy could not work in our present condition due to the fallen state of humanity. Only when a ruler is capable of being truly righteous and free from the temptations that pervert humankind can a theocracy function in a society. Our present form of government must take into account people's natural predilection to be corrupt and tyrannical.

Power must be decentralized and kept in check. The Islamic concept of an autocratic theocracy is therefore doomed to become a despotic tyranny, without the Son of God at the helm. This we have observed already, even at the level of single nations in those countries that have become Islamized.

Furthermore, the absolute standards of morality and justice set forth in the Word of God are, in fact, indispensable to achieve true justice in any form of government. And its exclusion from law is, without equivocation, the eventual dire exploitation of the masses for the benefit of the elite in power. It is the surest recipe for tyranny.

Sadly, I cannot say with confidence that those who died in our War of Independence and in the Civil War that followed have shed their blood for a cause that has completely triumphed.

The American experiment, our American republic, still hangs in the balance. The War of Independence may be over, but the Revolutionary War is not.

Abraham Lincoln was murdered so his dream could not be completely fulfilled. Today, we can no longer afford to think provincially. Our human civilization has advanced in technology to such an extent that the oceans that once kept us away from the turmoil of others are no longer adequate to isolate and protect us. But we must walk circumspectly and not allow the globalists to undermine our national distinctive. We must not allow our unique revolution to be sacrificed to the promoters of the globalization of our planet. The centralization of power has always been, is, and ever shall be the shortest route to tyranny.

The globalist ideology of borderless nations is nothing more than an insidious ruse to bring a global charter above our Bill of Rights and constitutional protections. If we hope to oppose this tyranny, we must understand how to intelligently oppose this view that has shrouded people's minds. We must educate ourselves to intelligently communicate the truth and expose the great evil of this worldview. That requires some measure of effort and time from each of us. To vacillate is to capitulate the future to the Enemy of Man.

Absolute Law vs. Popular Law and Sharia Law

Lies do not become truths, evil does not become good, and the slavery of the weak and powerless does not become morally acceptable simply because the majority wills it so or because the powerful elite demand it. Neither can reason alone give humans the anchor for truth. Each person may manipulate reason into rationalization to mask the selfishness and wickedness that people are prone to have. Without God as the final arbiter of truth and justice, there can be no absolute truth or justice. Without God, all human views are relativistic.

In order to implement law as king, there is the undeniable need to begin from the substratum of absolute truth as revealed in the scriptures. Anything else would simply resolve in shifting standards that would invariably end with the strong preying on the weak. And the law would invariably change until it met the needs and greed of the elite at the cost of the weak.

Relativism will therefore invariably lead back to the very thing we so abhorred and what fueled our American Revolution; that is, the unjust notion that the king is law. Whether that king hides behind an oligarchy or a tyrant or a central committee is immaterial. The final result has always been, is, and always will be the loss of personal liberty for the common people.

It is for this reason that we should carefully consider the integrity of our leaders. It is not eloquence or charisma or good looks that make a leader good. Beware of people who hunger for power, no matter how charismatic they may be. Those who covet power are the least capable of wielding it justly. Beware of the narcissist who wears his or her badge of arrogance as strength. Strength it is not. Weakness it is. These are the people the Enemy of Man can best manipulate to accomplish his evil machinations in our earth. It is the humble that God lifts up; it is the proud that He resists.

Every nation on earth has eventually self-destructed because it abandoned the moral standards given to them by God and ventured into the bottomless canyon of selfishness. When selfishness becomes acceptable to a culture, the death of that culture is inevitable as its citizens elect those who, like them, are selfish. In a very short time, rulers become ever more empowered by the Enemy of Man to enslave the masses.

The proponents of relativism point to Islam as promoters of absolute law in order to discredit the idea of absolute law. We have already pointed out that Islam adheres to a doctrine of abrogation, which, in essence, makes its laws subject to the will of the prophet of the day. That is not really absolute law.

Furthermore, Islam's laws are not based on the revelation of God found in the Hebrew Scriptures. There is but one God. There is but one source for absolute law. All spokes may lead to a hub, but all religions do not lead to God. All religions may contain some elements of truth, but wherever they differ from the Hebrew Scriptures, they are wrong. The postmodernist interpretation of religion that espouses this syncretic approach to truth is not any more Christian than Islam or Wicca. Moreover, Islamic Sharia law forces on humanity the rule of an autocrat as the final authority on all matters. It is a return to the concept of king is law, only dressed in religious terminology. The caliphate is, in essence, the emperor. It is no different from the papal throne of old that also sought to rule all humankind, in direct opposition to the judgment of the Tower of Babel that sought to decentralize power on our planet.

Correspondingly, freedom should be secured for all individuals to worship according to their consciences, as long as their worship does not entail criminal activity. Here and only here is true tolerance. For if the foundation of law becomes, instead, the arbitrary relativistic standards of naturalism or occultism or syncretic pantheistic faith or the autocratic rule of a religious imam, then all freedom for those with another perspective will disappear, and justice for all with it.

The notion that all people are created equal and are therefore entitled to certain inalienable rights stems directly and exclusively from the Judeo-Christian worldview. That differentiates the Judeo-Christian worldview from the Muslim idea that humanity is divided into two categories: Dar al-Islam (the Land of Peace or Submission) and Dar al-Harb (the Land of War).

Those who are Muslims belong to the Land of Peace and are declared to be worthy. Those who are not Muslims are infidels and have no intrinsic worth. The killing of an infidel is, in fact, the religious duty of the Muslim to expedite the establishment of the kingdom of the Mahdi. It is the embodiment of intolerance and tyranny in the political and spiritual arenas.

Muslims believe that in the day the caliphate is declared in Mecca, Islam will dominate the world, and the Mahdi will cut off the heads of all infidels who do not bow their knees to Allah. For that reason, all Muslim nations that have implemented Sharia law are autocratic tyrannies that do not provide freedom for the individual to express any view that is contrary to their theological construct as determined by their clerics. It is illegal to build a church or even bring a Bible into Saudi Arabia. Those countries conquered by Islamic terrorists not only destroyed any churches that existed there, even before Islam was born, but attempted to eradicate all Christians from their land, be they women, men, or children.

The Judeo-Christian worldview alone practices true tolerance, believing that a person's faith is a matter of private conscience. Muslims, on the other hand, believe that all aspects of life must conform to Sharia law traditions established by their clergy. This extreme that seeks to convert all nations to Sharia law by the edge of the sword will serve only to foment violence and justify Satan's tactic by discrediting monotheism. This ammunition used by the Western progressivist-socialist-collectivist movement quite deftly declares all monotheistic views as divisive and conflictive. Christians and Jews will be equally thrown into the same ditch.

In this way, the non-Islamic nations of the world will be led into the acceptance of a global union to defend against the Islamic threat under either an atheistic or pantheistic and syncretic view, which they will declare as the only rational path to global peace and religious tolerance. But beware of this trap. This will also never lead to true tolerance.

I am, however, of the opinion that Western globalists will underestimate the demonic power of Islam. As I see the prophecies of the scriptures unfolding, it has become quite evident to me that the Antichrist will quite likely be the Mahdi. The establishment of the caliphate by ISIS is not yet the caliphate that can be accepted by all Muslims, for the Madhi must be declared so in Mecca. That has not yet happened. And although this is a subject matter for another book, I suspect that the Mahdi will rise from Turkey.

Even if Western globalists succeed for a brief period, it will never result in true peace, for they will never be tolerant of any who espouse absolute standards. The shifting sands of pantheism and the enlightenment ideology will never produce freedom. It will inevitably lead to tyranny and the rule of the powerful. All these false religious ideologies lead to tyrannical political systems.

Law must be derived from the basic moral standards revealed by the creator and known to every culture that developed immediately after the Great Flood. These are also held in common by nations that have been influenced by Jews and Christians. Moreover, these fundamental moral precepts have been historically and generally accepted by every other nation, regardless of their religious background, as a faint memory from their inception. This is so because the religion of Noah was at the root of all the fledgling nations.

This universal nature reflects our common heritage since after the Great Flood, we were one people with a single language and a common religion. They are universally regarded as the basic notions of moral behavior and provide the substratum for the universal acknowledgment of the fundamental rights of humans.

However, their authority rests not on their universal acceptance but on the fact that they were drawn from the divine revelation of the creator. This is the only foundation that can lift these precepts above the relativistic quagmire of private opinions.

Paramount in this universal law is the very foundation of human civilization—the sanctity of all human life. The corruption of these standards is a historical reality that evidences the sin nature of humans. But none can argue that the kernel of truth exists in every corner of the world and that it existed with more clarity at the inception of their civilization. These form the basic framework of moral conduct that establishes the basis for equal justice for all and elevates the value of human life and acknowledges our inherent right to be free.

We may have clear access to these standards through the miracle of the revelation of God and the miraculous preservation of its contents throughout our history. Within the Holy Scriptures is prima facie evidence of its divine mandate. For this reason, our Founding

Fathers placed the Ten Commandments in the very building of the United States Supreme Court.

How long our new progressivist-controlled government will allow the Ten Commandments to stay there is looking pretty grim. On October 6, 2016, after a protracted legal battle, an Oklahoma court forced the removal of the Ten Commandments from the state's capitol. Then the once 5-4 majority of constitutional conservative justices on the U.S. Supreme Court vanished with the mysterious death of Anthony Scalia. Fortunately, the miraculous election of Donald Trump in 2016 resulted in the appointment of Justice Neil Gorsuch, another constitutional conservative. Trump's subsequent appointment of Justice Brett Kavanaugh again brought us to a conservative majority. But the potential to appoint two more Supreme Court justices still exists, which will secure the future of our Constitution and subsequently our liberties. However, the deep swamp is hard at work to resist any efforts to roll back the progressive agenda, and the Democratic Party is racing to the left further and further every year and looking for any excuse to impeach President Trump and prevent all this from happening.

This battle to protect the Constitution from those progressives who want to rid themselves of the shackles created by the separation of powers is not new. It began with Woodrow Wilson, the second progressive president of the United States. As Glenn Beck explains in *Liars*,

> Perhaps most damning of all, Wilson is the father of the single biggest philosophical threat to the Constitution the country has ever faced, one that splits the U.S. Supreme Court to this day. Wilson believed, in violation of everything the Founders stood for, in a "living and breathing Constitution" that "can and should be modified by its environment." These "living political constitutions must be Darwinian in structure and in practice," he wrote in 1908.

The Constitution was, in Wilson's mind, subject to the concepts of "survival of the fittest"—not bedrock at all but more like shifting sands. In the Wilsonian view, the government had to keep evolving, changing to meet the needs of the environment around it. There was no such thing as natural rights endowed to us by our creator or immutable principles.

Wilson's scorn for the Constitution rings clear and strong. He even derided the U.S. system of checks and balances: "No living thing can have its organs offset against each other, as checks, and live." He ridiculed the idea of individual rights.[17]

The undoing of the American Revolution is the primary goal of progressivists who want to amass central power in order to carry out their social policies of the redistribution of wealth by force and to undo all the individual liberties given to us by our creator and protected by the Bill of Rights. The reinterpretation of the U.S. Constitution through the Darwinian-atheistic-relativistic grid will erase all obstacles for the creation of an all-powerful federal government that will rule with an iron fist, as we have witnessed repeatedly in all nations that have accepted the collectivist worldview.

The battle between the Judeo-Christian worldview held by our Founding Fathers and the Darwinian-atheistic-progressive view is coming to a climax in our time. The fate of the American Revolution hangs in the balance, and the fate of our children go with it. Justice and liberty for the common person dances upon the edge of a blade, and what we do or fail to do may tip it one way or the other.

The Creator Alone Is the Ultimate Measure of Justice

The atheistic-Darwinian ideology of the evolution of truth cannot and will not provide the substratum for individual liberties. It is the tool of tyrants to control the masses and make the state god.

All rights are then arbitrary. Without God, there is no such thing as justice. There is only what is pragmatic for the government and the elite in power.

But there is a God, even if we have forsaken Him. His decrees are absolute and just. And it must be understood that without His revealed propositional truth in the Holy Scriptures, we would have no standard from which to measure truth or justice or from which to establish a just form of society. He is the source of truth, and His truth is eternal because the Lord changes not. "For I, the LORD, do not change" (Mal. 3:6).

This is the fundamental crux of our Judeo-Christian culture. Nevertheless, such axiomatic notions are universally understood and accepted by all people to be our fundamental human rights—the rights of life, liberty, and the pursuit of happiness. And for this reason, government must be limited to protect the individual from the abuse of power that is the natural disposition of humankind in our fallen state.

Our government should be limited to prevent it from taking these basic freedoms from us. The purpose of human government is to ensure that people are free. The purpose of our Constitution is to cement those individual rights for perpetuity. The role of government is not to dictate our lives but rather to get out of the way and allow us to live our lives. Its purpose is to ensure these freedoms:

1. To express oneself either by voice or print or any other medium in the public square as well as in educational institutions
2. To have equal access to opportunity
3. To have the right to own property
4. To have the right to worship freely according to our conscience and be able to express that without censorship
5. To be considered innocent until proven guilty
6. To be judged fairly by a jury of our peers
7. To be free from tyrannical coercion and not be arrested without due process of law

8. To have equal access to education and opportunity
9. To ensure that all common citizens are not exploited by the ruling or wealthy class
10. To protect small businesses from unjust and criminal abuse by large businesses
11. To be governed by a just representative form of government that derives its powers from the consent of the governed
12. To be free to travel
13. To have the right to bear arms for our own defense and to secure our liberty from a federal government gone rogue
14. To dwell in relative peace and safety, providing a safe environment for our families
15. To be able to vote for those leaders who represent our own personal views

Our individual rights cannot be enjoyed without the enforcement of certain legal restraints that outlaw immoral behaviors such as murder and theft. Our liberties end where the nose of our neighbor begins. All freedoms must have boundaries to protect them. Differences can be maintained and protected as long as they do not infringe on the liberties of others. Our freedom to worship and observe these minor differences should not be curtailed within our respective communities as long as they do not transgress these basic and elemental laws sifted from the propositional revelation of God and enumerated in our U.S. Constitution.

Here, and only here, do we find true tolerance in any given society. Such elemental rights could never evolve from a pantheistic, polytheistic, atheistic, or even Muslim worldview.

The pantheists insist that anyone who espouses an absolute concept of truth and morals must abandon this view in favor of relativism and the synchronization or fusion of all similarly related syncretic religious views. The Muslims insist that all humanity must be forced to conform to Sharia law, which is a narrow religio-political system that does not recognize the right of the individual to have freedom of expression or the right to believe in another religion other than Islam.

In either of these extremes, true tolerance is cast aside, and its eventual consequence will be war and violence toward those who do not accept their self-limiting view of reality. It is precisely that balance that our Founding Fathers attempted to reach when they formed the United States of America based on their underlying Judeo-Christian worldview. And it must be understood that all European nations owe their concepts of jurisprudence to the same Judeo-Christian worldview that shaped their formation.

True justice is not subject to a popularity vote. It is an absolute set forth from the bedrock of the revealed mind of God in His scriptures. Therefore, the desires of the majority cannot infringe upon the fundamental rights—the God-given rights—of the individual or the minority. And the collective is best cared for when individuals are free within the peripheral confines of this absolute law. True freedom cannot thrive without these limited and limiting boundaries.

There are boundaries to freedom that God has designed in order to prevent anarchy and chaos. Those boundaries limit a person's absolute freedom, but those boundaries must only limit those things that transgress God's revealed absolute moral dictums. Moreover, those boundaries must be constant and secure, not subject to the whims of popularity or of the power elite. Nor should they be partial to any particular subsection of our society.

An absolute law equally protects every human being, regardless of their ethnic background, religion, age, social status, gender, or any other descriptive component. Neither pantheism nor any other religion than the Judeo-Christian view could have spawned the American Revolution. No religion can establish an absolute standard for morals and justice if there is no final resting point for truth, morals, justice, or the transcendental value of human beings. This paramount truth is little understood by most.

Our unique form of government was made possible only because it began upon true biblical principles that ensure the basic dignity of every human being. Therefore, we have a foundation from which we can determine that the paradigm of reality is not the survival of the fittest, which can never hope to ensure social justice. All humankind has been equally created with special significance, which entitles each and every person to equal freedom and justice.

A relativistic foundation can only lead to the powerful tyrannizing the weak. Understanding that is crucial if we are to continue in the American Revolution. If we fail to understand this, our glorious revolution will come to a grinding halt, and our nation will become another run-of-the-mill tyrannical form of government.

We must never move from the bedrock principle of our American Revolution that we are endowed by our creator with these inalienable rights! Here, and only here, is the basis for true freedom of any kind. Especially important to our Founding Fathers was freedom of religion and the limiting boundary that this freedom cannot encroach on another's individual freedom to exist.

Religious traditions such as the outrageous religious practice of human sacrifice promoted by the enemy of humanity could not be a protected religious freedom since it interferes with the personal freedom of the individual being sacrificed. And for the very same reason, the Islamic ideology that legitimizes the murder of individuals who do not hold to Islam's religious creed and that artificially brands them as a subhuman category—infidels—is equally illegitimate in our constitutional republic.

Without this Judeo-Christian substratum, there can be no basis for real, tangible religious freedom for all religious views; there can be no true tolerance. And equally so, there can be no other adequate substratum to ensure the basic freedoms of the individual.

Naturalists, pantheists, and Muslims find religious freedom in our nation, but Christians and Jews are not afforded the same freedom in nations founded on other religions, whether theistic or anti-theistic.

The freedoms we offer in America cannot be confused with seditious religious beliefs that are intent on destroying our Constitution. In the name of religious freedom, Muslims are teaching on our American soil that the Constitution is an illegitimate document and that it must be overturned and replaced by Sharia law. That should not be allowed, and it is nothing less than sedition. Sharia law cannot be included as a lawful religious doctrine that is tolerated in our nation. Anyone who wants to reside in our nation must be willing to pledge their allegiance to our Constitution and cannot be allowed to teach sedition under the guise of religious freedom.

It is more than ironic that Christians are prohibited from building churches in Saudi Arabia, while the Saudis have built hundreds of mosques in America in order to promote their religious worldview in our nation. And they do not even allow non-Muslims to step into the city of Mecca. It is illegal for a non-Muslim to even drive through that city.

In the name of religious tolerance, Muslims have been allowed to build some 35 military compounds on American soil under Sheikh Gilani and the auspices of The Muslims of America. These compounds are used to train jihadists in such religious duties as making explosives, using guns to kill infidels, learning the proper way to hijack vehicles and grab hostages, and disseminating seditious literature that calls for the destruction of the American government and the establishment of Sharia law in Washington, DC. This "oversight" is not an expression of religious tolerance. It is plain and simple stupidity. This must stop. Any group, religious or otherwise, planning to destroy our constitutional way of life must be immediately deported, never to be allowed reentry. To do anything less is to help them build their Trojan Horse.

But naturalism is not any better. The fallacious illusion that relativism could produce social justice is made evident in those countries where their religion is naturalism—countries such as China, North Korea, and the former Soviet Union. We find that Muslims, Jews, and Christians are inevitably equally persecuted there. This has been widely documented, and I need not take time here to enumerate the many resulting infractions against the dignity of humanity perpetrated in those nations.

Ironically, it is commonly believed in the West by naïve adherents of Western-style Eastern mysticism that Hinduism and Buddhism are tolerant religions. The truth is that Christianity is under severe persecution in India, even in our present time. Churches are being burned, pastors are being killed or severely beaten with sticks, assemblies are forbidden, and violence toward anyone who promotes another religion is quite prevalent. These things are happening in India as well as other countries in the Far East that espouse a pantheistic religion. True tolerance exists only where the Judeo-Christian worldview flourishes. This is an undeniable reality that no one can dispute.

Notes

1. John Whitehead, *The Separation Illusion* (Milford, MI: Mott Media, 1977), 26.
2. Paul Rodriguez, "God's Microscope," unpublished paper.
3. U.S. Constitution – First Amendment, Legal Information Institute, Cornell Law School, https://www.law.cornell.edu/constitution/first_amendment.
4. George Washington, "Thanksgiving Proclamation," October 3, 1789, Library of Congress, https://www.loc.gov/resource/mgw8a.124/?q=1789+Thanksgiving&sp=132&st=text.
5. Ibid.
6. Ibid.
7. "Proclamation Appointing a Day of Thanksgiving and Prayer, 11 November 1779," National Archives, https://founders.archives.gov/documents/Jefferson/01-03-02-0187.
8. Ibid.
9. Ibid.

10. "From the Danbury Baptist Association," October 7, 1801, The Papers of Thomas Jefferson, 35:1 (Princeton, NJ: Princeton University Press, 2008), https://jeffersonpapers.princeton.edu/selected-documents/danbury-baptist-association.

11. "Thomas Jefferson's Letter to the Danbury Baptists," January 1, 1802, Library of Congress, https://www.loc.gov/loc/lcib/9806/danpre.html.

12. Ibid.

13. Francis Schaeffer, *A Christian Manifesto* (Westchester, IL: Crossway Books, 1982), 38.

14. "Thomas Jefferson's Letter to the Danbury Baptists."

15. National Archives, "Declaration of Independence."

16. Francis Schaeffer, *The Great Evangelical Disaster* (Westchester, IL: Crossway Books, 1984), 141.

17. Glenn Beck, *Liars: How Progressives Exploit Our Fears for Power and Control* (New York: Threshold Editions/Mercury Radio Arts, 2016), 58.

CHAPTER 5

● ● ●

THE CORRUPTION OF NATIONALISM

*The God who made the world and all things in it, since He is Lord of heaven and earth, does not dwell in temples made with hands; nor is He served by human hands, as though He needed anything, since He Himself gives to all people life and breath and all things; and **He made from one man every nation of mankind** to live on all the face of the earth, having determined their appointed times and the boundaries of their habitation, **that they would seek God**, if perhaps they might grope for Him and find Him, though He is not far from each one of us* (emphasis added).

—Acts 17:24–27

God made from one people every nation of humankind that we might seek after Him. He chose to divide humanity into separate nations in order to protect us from the natural repercussion inherent in a centralized global government; that is, a repressive tyrannical system.

The centralization of power has ever been the nemesis of individual freedoms. No individual freedom can exist without a limited government. No government can ever be limited if the economic power of a nation is centralized and monopolized, for human nature is such that it will eventually seek inordinate political power when economic power enables it to do so. These truths are absolutely axiomatic.

In every epoch of our human history, an ancient cosmic battle has raged as the demonic hierarchy has manipulated people who ever craved for power and are devoured by greed in their ongoing plan on earth to establish a global kingdom. Each time that Satan has almost reached his global ambition, God has countered and foiled his plans.

Toward the end of the first earth, Satan had almost reached his global aim, and then God destroyed the earth and started again with Noah. The second earth began as one people with one language living at first in tribal units. Not long after, as the numbers grew, the people coalesced into one governmental power that resided in the Valley of Shinar.

Within a few hundred years after Noah disembarked from the ark, humankind was on the brink of falling prey to Lucifer's demonic plans again as the bloody religion of the serpent once again took power. And so God divided humanity by confounding the languages. In doing so, He once again foiled the global aspirations of the demonic hierarchy through the judgment of the Tower of Babel.

In every epoch since, the drive to establish an international government and a global monetary system has been engineered by the demonic forces seeking to undo the judgment of the Tower of Babel. For that reason, true and balanced nationalism has ever been under attack. But God instituted the separation of humanity into nations in order to protect humankind from the tyranny that will inevitably come if we capitulate our divinely mandated separate national powers into a single global government.

God instituted human governments in independent nations in order to have authority over the people and provide a structure that would secure justice, liberty, and domestic tranquility. For that reason, the scriptures admonish us to submit ourselves for the Lord's sake to these various human institutions whose responsibility is to punish evildoers and reward those who do right (1 Pet. 2:13–14).

It is not the responsibility of the government to care for the people. It is the responsibility of the government to protect the people and secure the personal freedoms necessary for them to care for themselves. Our collective responsibility applies only to care for those who cannot care for themselves. Those who can work must care for themselves.

Unfortunately, all human institutions have inevitably become corrupted. They were intended to protect human liberties, defend people from the exploitation of the powerful, and create domestic tranquility. But human nature being what it is, all human governments have inevitably become oppressive systems that have enslaved their constituents.

The very nature of our human condition inexorably leads these institutions to become repressive tyrannies. Either the rulers become corrupt or the political system was designed for the exploitation of the masses to begin with. In every historical case, they have all rebelled against God's design for a just form of human government. Consequently, these false ideologies contrived by humans and spurred by demons who corrupt God's intention for the nations, have inevitably led to the exploitation of the masses.

It must be stated that the corruption of the one true religion passed down by Noah to his descendants contributed greatly to the corruption of the political ideologies established by humankind.

People filled with greed for power and wealth have historically forced these aberrant forms of government upon people through armed aggression and military conquest, many even demanding to be worshipped as gods. Do not think that this is unique only to the ancient past. We can point to North Korea in the present and Japan in World War II.

All these humanly contrived false ideologies that pervert true nationalism stem either from the misunderstanding of our individual identity in relation to God or from the wanton greed and selfishness of tyrants. Each of these is constructed with a corrupted view of our true identity as human beings. They purposefully deny humans their proper place as beings with infinite intrinsic transcendental value because we have been created in the image of God; they heartlessly use people to expand their empires to loot and pillage what other people have lawfully labored to acquire.

It is because we are created in God's image that our individual rights are inviolable. And for this reason, the nations were ordained to defend and promote those individual rights granted by God to each human being. If this fundamental foundation is either removed or ignored, then government becomes god, and each of us becomes a slave to the government to be used for its own greedy purposes.

This travesty of our inherent human rights began with monarchies when kings filled with insolence and pride made themselves gods, usurping the place of the creator. These kings abandoned their rightful place as defenders of the individual rights of their people and instead enslaved them and used them to amass their wealth and power. War became the norm in human history as people found it faster and more lucrative to loot the wealth of others than amass it through honest labor.

All false political ideologies begin with a departure from their initial intentions to create human governments as dictated by a just and loving God. They may differ in outward form, but they have the same rebellious foundation against the order that God has designed for humankind, inevitably leading to tyranny, oppression, and subjugation.

Rebellion against God's order by those who govern always leads to tyranny. Rebellion by those who are governed leads to chaos and anarchy. Anarchism at one extreme repudiates any extraneous authority over each person and falsely elevates each person to a position of godhood. The human will supersedes all others, and humans become a law unto themselves. This also leads to violence and savagery that inevitably allow a strong person to emerge and bring order out of chaos by the power of the sword. Unfortunately, when such chaos creates such lawlessness, people are prone to sacrifice their freedoms for security. But the security of tyrants is ever an illusion.

The rule of a benevolent monarch may create a just social system of governing, but historically, that has always been the exception to the rule. The fundamental problem is that the centralization of power in one individual provides too easy a target for the Enemy of Man to pervert and corrupt. It is our fallen nature that looms largely in our history and evidences the endless appetite of greedy people for power and wealth.

In monarchies that have gone awry, the king is elevated to the position of godhood. His word becomes the law of the land and is therefore subject to his manipulation for selfish reasons. Rare have been the monarchies that have had benevolent rulers and provided a just form of government for the people. The centralization of economic and political power has been the enemy of the common person. Whether in the form of an oligarchy, a monarchy, an autocratic despot, or a central committee, the liberty of the masses has suffered under their repressive yokes that used people as expendable assets to amass their wealth and power.

Each is a form of hyperindividualism that elevates either a single person or an elite group of the wealthy beyond the boundary that God has designed. All tyrannies elevate either one person or an elect few to a position of supreme power at the expense of the rest. It is an elitist individualism that devalues all others as slaves to these masters. These extremes are all at the opposite end of God's intended governmental system. Each fails to find the proper balance that God mandated for all humankind.

Each of these tyrannical systems of governing shares the same foundation of the centralization of economic power and political power. The rise of one always leads to the rise of the other. Sometimes, the centralization of economic power rises first and then takes over the political power, as is happening in America. And sometimes, it is the centralization of political power that leads to the centralization of economic power, as it happened in the Soviet Union, China, Cuba, Venezuela, and Iran. Either way, the inevitable result is always the loss of freedom and the economic enslavement of the masses.

The centralization of economic and political power has been achieved historically through these three major venues:

1. Monarchies – an elitist aristocracy that traditionally owns most of the land and provides for its citizens a modicum of protection from outside invaders but at the price of being enslaved to the monarch.
2. Collectivism – an elitist party bureaucracy that rewards an elite few and in varying degrees of ruthlessness stamps out any dissenting views.
3. Supernationalism – expressed in three forms:
    a. through a religious autocrat such as championed by Islam.
    b. through an autocratic despot or dictator such as Mussolini or Hitler and thousands of other tyrants who have risen to power in nations throughout the earth.
    c. through an oligarchy or plutocracy, the rule of a nation through an elite wealthy group whose economic power is absolute and able behind the scenes to control a country through a shadow government.

Oligarchies arose in most Western nations around the turn of the twentieth century. It is my opinion that America also effectively became a fledgling oligarchy in 1914, and its political power has increased during the last century to an alarming degree.

Due to our unique structure of the separation of powers and our magnificent Constitution, a complete grip on our nation has not developed as quickly as it has in other less fortunate nations. The ultimate goal of all oligarchies has always been to eradicate the middle class in order to have no economic competition and to keep them from amassing the finances to create any credible political opposition to their particular agendas.

It is my opinion that if our young people are not awakened to the dangers of socialism, we will soon reach the checkmate move that will effectively erase the separation of powers and the protection of our Constitution. The clever ruse of oligarchies is to claim that they are championing the working person against the wealthy elite when, in fact, the wealthy elite are the real movers behind the socialist policies. Bankers know well that in a more centralized government, they have more control and that the more debt a nation acquires, the more interest they will collect on that debt. It is the fastest and most efficient way to fleece the flock. It will spell the end of the great American middle class that has resulted in more political freedom, personal liberty, and affluence than any other nation in the world.

Diagram 1: Relationship between Nationalism, Anarchy, Monarchies, Supernationalism, and Collectivism

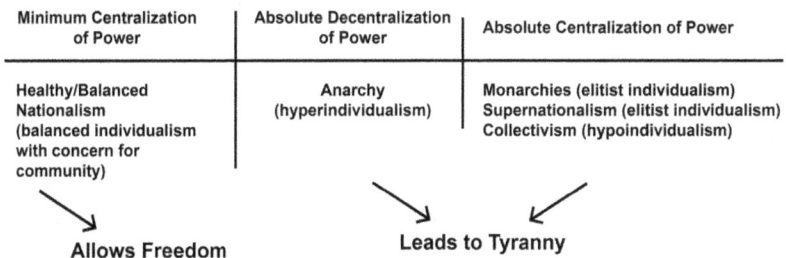

Minimum Centralization of Power	Absolute Decentralization of Power	Absolute Centralization of Power
Healthy/Balanced Nationalism (balanced individualism with concern for community)	Anarchy (hyperindividualism)	Monarchies (elitist individualism) Supernationalism (elitist individualism) Collectivism (hypoindividualism)

Allows Freedom   **Leads to Tyranny**

Anarchy, Collectivism, and Supernationalism Inevitably Lead to Tyranny

The anarchist champions the complete or absolute decentralization of power. And power resides only in the individual. Therefore, it seems to be superficially at the opposite extreme of collectivism and supernationalism, which champion the centralization of economic and political power. But as history has taught us, in the end, anarchy, collectivism, and supercapitalism always lead to the same tyrannical system that enslaves the working person.

At the same end of the spectrum within these false ideologies, anarchism champions an egotistical unbalanced sense of hyperindividualism that leads to the natural and inevitable consequence of chaos and violence through the subsequent struggle that results from the survival of the fittest. The absolute decentralization of power inevitably evolves to a despotic tyranny from the helter-skelter of the milieu as the public, desperate to bring order to the resulting chaos, blindly sacrifices their liberties for security.

With the advent of the Renaissance or Enlightenment worldview by a vast swath of our Western culture, which categorically rejects the creator, many in our modern times have become proponents of the anarchist ideology. It is the natural consequence of the occult false promise that we can be our own gods, and thus the call of the occult for every person to rise and reclaim their natural birthright: godhood.

Therefore, anarchism is the by-product of a hyperindividualistic worldview that rejects all forms of authority as illegitimate. It is the very essence of the rebellious spirit in the great deceiver who brought about the heavenly insurrection. It is the natural reflection of an extreme hyperindividualism that perverts the Judeo-Christian concept of the *Imago Dei.*

The magnificent lie brings with it the ideology that each of us is absolutely free to choose our own set of morals and ethical standards. But through the implementation of this ideology, no society can ever hope to function. Such a system will in every case lead to such chaos that it will, in the end, invariably promote the rise of a tyrannical despot who will rule all others for his or her own selfish ambitions. We can point to the French Revolution as a historical example in the second earth of this inevitable process. It was exploited by Napoleon to establish his autocratic empire. Today, there is some resurgence in the popularity of this ideology in conjunction with the rise in popularity of the occult, especially in Europe. We do not have to travel far to find the symbol of anarchism—the letter A—painted in red as graffiti on posters and the walls of buildings and subways in major cities. For the most part, the majority of people recognize this ideology as untenable.

But our foe is clever beyond our wildest dreams. He knows how to redirect his energy to accomplish the same goals another way. Although the preferred ideology for those in the occult is anarchy, he well knows he must deceive the masses. For this reason, supernationalism and collectivism have been his most effective weapons of choice in the second earth as he continues his plan to set up a global throne.

At the same end of the spectrum, where power is absolutely centralized, we find both supernationalism and communism. Supernationalism also champions an elitist form of individualism that leads to the natural consequence of war through the conquest and subjugation of those deemed inferior. Examples of such supernationalist aspirations can be seen in Nazi Germany, Benito Mussolini's Italy, the Japan of World War II, the Ku Klux Klan (KKK) and other white supremacists in the United States, and Islam.

These have been incorrectly labeled as opposites. The popular modern misconception that labels communism as a leftist ideology and Nazism as a right-wing ideology is completely fallacious. They are not at opposite poles. They are all imperialist-colonialist-globalist ideologies that seek to subjugate all other nations because they have been deluded into thinking that either their nation or their race or their economic system is superior to all others.

In the end, all of them promote globalism and the centralization of economic and political power. The Nazi Party is, after all, the Nationalist Socialist Party. It is a socialist system. They are all corruptions of a healthy and balanced form of nationalism that invariably leads to oppression, war, tyranny, and globalism.

The supernationalism advocated by the Nazi worldview brings us to the same global tyranny that collectivism and Islamic colonialism champion. Without a doubt, humanity has witnessed for too long the calamitous aftermath of supernationalistic aggression. Nations have been whipped into a blood-frenzy by wicked monsters of greed and power in order to satiate their desires for glory and material plunder of their own brothers and sisters. We are all the children of Adam and Eve.

How many young men and women have been senselessly sacrificed upon their abominable altars of conquest and greed? Only God knows. Those who champion the collectivist ideology are eager to use the atrocities of supernationalism as a way to discredit a balanced form of nationalism. But collectivism is also a corruption of nationalism that equally leads to tyranny.

Here in America, we can see the very same violent and intolerant tactics in Antifa groups that seek to violently stop all opposing worldviews and promote a collectivist ideology that sacrifices individual freedoms and rights for the tyranny of the collective.

It is an indictment on our American universities that they have spawned this malevolent and exclusivist ideology that seeks to oppress all opposing views through violence, oppression, and intimidation. They are the executioners of true free speech and the enemies of our First Amendment rights so hallowed by our Founding Fathers as a direct result of their Judeo-Christian worldview. It is my opinion that the federal government should deny all federal aid to educational institutions that do not ensure freedom of speech to all their students.

No honest person can deny the horrors brought upon humanity by supernationalism. Both collectivism and supernationalism are the enemies of true nationalism and the freedom of speech and religion espoused by our American Constitution. We must not allow the promoters of either collectivism or supernationalism to destroy the proper and balanced sense of true nationalism in our country. This is not a false, prideful supernationalism that looks down on other cultures as inferior. It is rather one that recognizes the uniqueness of each culture and the equality in value and uniqueness of each individual in every culture. It is equality in value with distinctions in cultural traditions. True nationalism is a natural extension of the God-given rights of individuals.

The Judgment of the Tower of Babel – Decentralization

It must be noted that God's statutes and moral precepts are the fundamental basis for liberty, while maintaining the proper respect for the individual within a governing structure. Such a fundamental truth was surely understood by Noah and his children. But in a relatively short period of time, it was ignored as their descendants strayed from the one true religion.

This concept of limited government, which is based on God's principles, is therefore crucial in maintaining our freedom as citizens of a nation. There are three basic principles that led our Founding Fathers to establish a form of government that was both representative of the will of the people and limited in power in order to forestall the inevitable abuse of political power.

1. The first principle is the sacred acknowledgment that our human liberties stem from God and not from any human institution. God created us in His image and therefore endowed us with infinite value. That is the sole foundation for individual rights that guided our Founding Fathers in the creation of this great American experiment and the Bill of Rights.

2. The second principle is the Judeo-Christian understanding of the fallenness of humankind and people's natural predilection

to abuse power when attained. This is the reason our Founding Fathers established the separation of powers to forestall the inevitable corruption that is common to all human endeavors throughout history.

3. The third principle is that in order for justice to exist, a nation must have an absolute moral standard from which to establish its laws. That is what motivated our Founding Fathers to create a constitutional republic in which law was king.

We should not ignore these three fundamental premises. Our failure to implement them and allowing globalism to replace our national sovereignty will result in the same disastrous consequences that brought about the judgment at the Tower of Babel. The battle is not between Republicans and Democrats; it is between those who want to maintain our national integrity with a limited federal government—one protected by our Constitution that birthed our liberties and financial success—and those who want to undo our national sovereignty with open borders and centralize control of every aspect of our lives by a massive federal bureaucracy.

Because of the judgment at the Tower of Babel, nations developed as a natural consequence of the dispersion created by the confounding of the ancient common language, which was probably Sumerian. God's purpose was primarily to decentralize power in order to keep Satan from attaining global dominion and also to protect human rights.

In other words, from a Judeo-Christian perspective, the decentralization of power was the tool mandated by God to keep Satan in check. The socialist-collectivist-progressivist ideology is the antithesis of this decentralization. It is no wonder Joseph Stalin said in a conversation with Winston Churchill in November 1943, "The Devil's on my side, he's a good Communist."[1]

It is logical to assume that these nations, after the dispersion at the Tower of Babel, were formed in order to provide a system of government whose power was centralized only to the minimal extent that it could allow them to govern the people with justice. It was God's intent that humankind would turn to Him while enjoying the order and protection afforded by a governing authority that could maintain order and peace in a community.

You might ask, *What is wrong with a unified people? Is God against unity?* Not at all. God desires for His children to be united in harmony. While it is true that immediately after the Great Flood we were one people, speaking only one language, and that this would have been the ideal governmental system for our planet since true unity and community pleases God, the sad fact is that it was our sinful condition that kept us from succeeding in this form of union.

Our human frailty proved to be the Achilles' heel of this form of government. This is the second fundamental premise understood by our Founding Fathers and the reason they took such care to limit the power of the federal government and separate the judicial, legislative, and executive branches.

The undeniable historical fact is that the centralization of power always results in the abuse of that power. It is for this reason that God divided us into separate nations. Had He not stepped in and divided us, Satan would have gained complete control of the global throne only a few hundred years or so after the Great Flood. How sad would that have been?

Therefore, God dispersed us throughout the planet and foiled Satan's plan for global dominion on earth. God consequently instituted national governing systems for the purpose of executing justice and protecting their people. Since that day, Satan has been hard at work to undo the work of God.

When again, due to our human condition, the patriarchal nations turned into monarchies, the Gentile nations once again failed to do God's will, turning to the religion of the serpent. God had to once more step in and intervene. The wholesale acceptance of the mystery religion of the serpent throughout all the Gentile nations brought with it wars, violence, human sacrifices, and untold injustices that marred the very purpose of their existence. The student of history must note that there is a direct correlation between the religion of the serpent and the corruption of nationalist governments.

It was then that God saw fit to establish a new nation from the loins of Abraham to fill the void left by the corrupted Gentile nations. For this purpose, Israel was separated from the other nations to be the light of truth to the Gentile world that had veered from its intended path. That is, the nation of Israel was created for the express purpose of championing truth and justice in a world that had turned its back on God.

> *And ye shall be unto me a kingdom of priests, and an holy nation. These are the words which thou shalt speak unto the children of Israel.*
> —Exod. 19:6 KJV

> *For thou art an holy people unto the LORD thy God: the LORD thy God hath chosen thee to be a special people unto himself, above all people that are upon the face of the earth.*
> —Deut. 7:6 KJV

Israel was not created as a monarchy at the beginning. But unfortunately, after some time, Israel also failed to fulfill its role as the bearer of truth and justice. The prophet Samuel warned them of the dangers of instituting a king and centralizing power, but they refused to listen. Samuel was distraught, knowing full well the consequences that would come through the centralization of power. But the Lord comforted Samuel, saying:

> *And the LORD told him: "Listen to all that the people are saying to you;* **it is not you they have rejected, but they have rejected me as their king***"* (emphasis added).
> —1 Sam. 8:7 NIV

History sadly records, as Samuel had feared, that a majority of the kings of Israel turned away from God and followed after the religion of the serpent. Time after time, the Lord sent His prophets to warn Israel of their impending doom if they did not turn from the religion of the serpent. Isaiah, Jeremiah, Amos, and many other prophets duly warned them that turning to the Baals would lead to death. But Israel did not listen and was taken out of the Promised Land.

We cannot help but hear the anguish in Jeremiah's words as he daily sought to turn his people from the coming calamity. Instead, the prophets who were sent to warn them were persecuted and killed. So, God instructed Jeremiah to tell Israel that they would be hurled from the Promised Land.

> *"Now when you tell this people all these words, they will say to you, 'For what reason has the LORD declared all this great calamity against us? And what is our iniquity, or what is our sin which we have committed against the LORD our God?' Then you are to say to them, 'It is because your forefathers have forsaken Me,' declares the LORD, 'and have followed other gods and served them and bowed down to them; but Me they have forsaken and have not kept My law. You too have done evil, even more than your forefathers; for behold, you are each one walking according to the stubbornness of his own evil heart, without listening to Me. So I will hurl you out of this land into the land which you have not known.'"*
>
> —Jer. 16:10–13

Is it not fair to say that this solemn warning may be repeated to our postmodern Western culture today? **"Behold, you are each one walking according to the stubbornness of his own evil heart, without listening to Me."** And yet we find today the very same question on the lips of those who mock His existence: "For what reason has the Lord declared all this great calamity against us?"

Closed are the eyes of the mockers, and futile is their darkened understanding. Their ears are stopped with wax, for they seek not truth but personal gain. Their reward will be striving after the wind and the vexation of their spirit. And so Israel was taken captive into Babylon for 70 years as prophesied by Jeremiah. The Temple that Solomon built was destroyed on the 9th of Av.

There, in Babylon, a few of the exiled Jews turned back unto the Lord. The people of Israel sought the face of God and recognized that God was visiting them. And God, true to His promise, released them from bondage, exactly 70 years later. The prophet Jeremiah had foreseen Israel's release, even before it had been taken captive.

> *"For thus says the LORD, 'When seventy years have been completed for Babylon, I will visit you and fulfill My good word to you, to bring you back to this place. For I know the plans that I have for you,' declares the Lord, 'plans for welfare and not for calamity to give you a future and a hope. Then you will call upon Me and come and pray to Me, and I will listen to you. You will seek Me and find Me when you search for Me with all your heart.'"*
>
> —Jer. 29:10–13

According to Babylonian chronicles, Nebuchadnezzar had taken some 109,000 people into exile from Israel. There, in Babylon, the Jews prospered, and many merchants grew rich. It therefore stands to reason that 70 years later, the number of exiles had increased significantly.

And yet only 50,000 had the courage and faith to return to the Promised Land and rebuild the Temple of the Lord. It seems that only a minority of people are ever willing to remain faithful to the truth. That is a universal truth that spans our entire human history.

But the Temple of God was rebuilt, even amidst great opposition, and God blessed Israel once again. And yet the people of Israel again strayed from the truth. And when their Messiah came to be their Passover Lamb, Israel did not recognize its Day of Visitation.

And so Israel was once more expelled from the Promised Land, and the second Temple was destroyed on the exact same date, the 9th of Av, so Israel would know it was a judgment from God.

Sadly, even the nation of Israel failed. Four hundred eighty-three years later (as predicted in Daniel 9:24–27) on the tenth day of Nisan (the day the Passover Lamb is presented for Passover), Jesus entered the Eastern Gate of Jerusalem riding on a colt, the foal of a donkey, which had been prophesied of old (Zech. 9:9).

As soon as He was approaching, near the descent of the Mount of Olives, the whole crowd of the disciples began to praise God joyfully with a loud voice for all the miracles which they had seen, shouting:

"BLESSED IS THE KING WHO COMES IN THE NAME OF THE LORD; Peace in heaven and glory in the highest!"

Some of the Pharisees in the crowd said to Him, "Teacher, rebuke Your disciples." But Jesus answered, "I tell you, if these become silent, the stones will cry out!"

When He approached Jerusalem, He saw the city and wept over it, saying, "If you had known in this day, even you, the things which made for peace! But now they have been hidden from your eyes. For the days will come upon you when your enemies will throw up a barricade against you, and surround you and hem you in on every side, and they will level you to the ground and your children within you, and they will not leave in you one stone upon another, because you did not recognize the time of your visitation."

—Luke 19:37–44

The Pharisees wanted Jesus to rebuke His disciples because they were ascribing to Him the praise that was to be given to the Messiah. They did not recognize the time of their visitation. And so we see again that only a minority stayed true to the Lord.

Therefore, God in His infinite wisdom saw fit to begin anew with the instrument of the church, composed of both Jew and Gentile. The Jews were not forsaken; they were an integral part of the church. As a matter of fact, Christianity views Gentiles as a wild olive branch grafted into the natural olive trunk of Israel:

> *But if some of the branches were broken off, and you, being a wild olive, were grafted in among them and became partaker with them of the rich root of the olive tree, do not be arrogant toward the branches; but if you are arrogant, remember that it is not you who supports the root, but the root supports you.*
>
> *And they also, if they do not continue in their unbelief, will be grafted in, for God is able to graft them in again. For if you were cut off from what is by nature a wild olive tree, and were grafted contrary to nature into a cultivated olive tree, how much more will these who are the natural branches be grafted into their own olive tree?*
>
> *For I do not want you, brethren, to be uninformed of this mystery—so that you will not be wise in your own estimation—that a partial hardening has happened to Israel until the fullness of the Gentiles has come in; and so all Israel will be saved; just as it is written,*
>
> *"The Deliverer will come from Zion,*
> *He will remove ungodliness from Jacob."*
> *"This is My covenant with them,*
> *When I take away their sins."*
>
> —Rom. 11:17–18, 23–27

Some very mistaken Christians have believed that the church has replaced Israel. True Christianity knows that God is not done with Israel. They will become the priests of God when the Lion of Judah comes to deliver our planet from the jaws of the Antichrist.

Yeshuah is simply Joseph, who was rejected by his brothers and given to the Gentiles for a period of time. The day will come when Israel will bow before Joseph in repentance. Their lives will depend on it when the seven years of famine strike the planet. Joseph will welcome them with open arms and deliver them. Elijah will return in that day to turn the hearts of Israel toward God. In that day, there will be a fountain of grace given to Israel when they recognize the Lion of Judah as the One who was pierced for our sins.

"Behold, I am going to make Jerusalem a cup that causes reeling to all the peoples around; and when the siege is against Jerusalem, it will also be against Judah. It will come about in that day that I will make Jerusalem a heavy stone for all the peoples; all who lift it will be severely injured. And all the nations of the earth will be gathered against it. In that day," declares the LORD, "I will strike every horse with bewilderment and his rider with madness. But I will watch over the house of Judah, while I strike every horse of the peoples with blindness. Then the clans of Judah will say in their hearts, 'A strong support for us are the inhabitants of Jerusalem through the LORD of hosts, their God.'

"In that day I will make the clans of Judah like a firepot among pieces of wood and a flaming torch among sheaves, so they will consume on the right hand and on the left all the surrounding peoples, while the inhabitants of Jerusalem again dwell on their own sites in Jerusalem. The LORD also will save the tents of Judah first, so that the glory of the house of David and the glory of the inhabitants of Jerusalem will not be magnified above Judah. In that day the LORD will

defend the inhabitants of Jerusalem, and the one who is feeble among them in that day will be like David, and the house of David will be like God, like the angel of the LORD before them. And in that day I will set about to destroy all the nations that come against Jerusalem."

—Zech. 12:2–9

In this passage written by the Hebrew prophet Zechariah, it is Jehovah who is speaking to Israel, warning them that the nations of the world will in that day be gathered against her to destroy the people of God. But in that day, God will instead destroy the nations of the world gathered at Megiddo. In the next verse, Jehovah continues His dialog and says He will pour out of the house of David and on the inhabitants of Jerusalem the Spirit of grace:

"I will pour out on the house of David and on the inhabitants of Jerusalem, the Spirit of grace and of supplication, so that they will look on Me whom they have pierced; and they will mourn for Him, as one mourns for an only son, and they will weep bitterly over Him like the bitter weeping over a firstborn."

—Zech. 12:10

When was Jehovah pierced? Why, if God is delivering them, will they be mourning? Isaiah saw that day:

Break forth, shout joyfully together,
You waste places of Jerusalem;
For the LORD has comforted His people,
He has redeemed Jerusalem.
The LORD has bared His holy arm
In the sight of all the nations,
That all the ends of the earth may see
The salvation of our God.

—Isa. 52:9–10

And so it shall be that the right arm of God with the nail-pierced scars will be bared before the eyes of Israel. Some 700 years before Christ, the Hebrew prophet Isaiah foresaw that the right arm of the Lord would be pierced, and he knew that many would not believe his message:

> *Who has believed our message?*
> *And to whom has the arm of the Lord been revealed?*
> *For He grew up before Him like a tender shoot,*
> *And like a root out of parched ground.*
> *He has no stately form or majesty*
> *That we should look upon Him,*
> *Nor appearance that we should be attracted to Him.*
> *He was despised and forsaken of men;*
> *A man of sorrows and acquainted with grief.*
> *And like one from whom men hide their face*
> *He was despised, and we did not esteem Him.*
>
> *Surely our griefs He Himself bore,*
> *And our sorrows He carried;*
> *Yet we ourselves esteemed Him stricken,*
> *Smitten of God, and afflicted.*
> *But **He was pierced through for our transgressions,***
> *He was crushed for our iniquities;*
> *The chastening for our well-being fell upon Him.*
> *And by His scourging we are healed.*
> *All of us like sheep have gone astray,*
> *Each of us has turned to his own way;*
> *But the LORD has caused the iniquity of us all*
> *To fall on Him* (emphasis added).
>
> —Isa. 53:1–6

He came first on the Passover to deliver us spiritually from the angel of death. By the shedding of His blood we are healed. By His sacrifice for our sins, we are forgiven. Those who outside of any works of righteousness they have done, by faith, brush the blood of the Passover Lamb over their doorposts will be saved from the Angel of Death. But He will return on Yom Kippur, not as the Lamb of God but as the Lion of Judah, to deliver us physically and establish the throne of David. If you are Jewish, then hear my plea: Do not fail to listen to Elijah, for he alone can turn you back to the truth. He alone can guide you to the Great Succoth that alone can save you from the coming terror of the King of the North.

He has sent us, both Jew and the Gentile believers grafted in the natural olive tree of Israel, to go throughout the nations to effect change in each of the nations. That is the task of the true Christian church composed of both Jew and Gentile—to transform our brothers and sisters from within through the gospel of the grace of God. Our responsibility is twofold. It is spiritual as well as physical. Make no mistake, our citizenship is really in heaven, but our earthly responsibility is to be the light on the hill and the salt of the earth, and to render unto Caesar that which is Caesar's in the nations in which God's providence has placed us. Our social responsibility is part of our spiritual duty. There is no dichotomy here.

Take hope, my Jewish brothers and sisters, God is not done with Israel, and its place in history is comparable to none other. Hear this, O Israel: Those who are truly Christians are your children in the faith. And we will not abandon you. Yours are the shoulders upon which righteousness will be exclaimed throughout the entire earth when the Messiah returns to sit on His global throne in Jerusalem.

Therefore, I urge you, my Jewish brothers and sisters, to take a stand now for righteousness, to be the light unto the Gentiles that you were intended to be. The King of the North will soon descend upon you to profane the sanctuary and take away the daily sacrifice. Your prophet Daniel saw this day. This is not a Christian prophecy.

THE CORRUPTION OF NATIONALISM

Your own prophets saw this day:

> At the appointed time, he shall return, and come towards the south . . . he shall even return and have an understanding with those who forsake the Holy Covenant. And arms shall stand up on his part, and they shall profane sanctuary and fortress, and shall take away the daily sacrifice, and they shall set up the abomination that makes desolate. **And such as do wickedly against the covenant shall be seduced by flatteries; but the people who know their God shall be strong and prevail.** And they who understand among the people shall instruct many; yet they shall fall by the sword, and by flame, by captivity, and by spoil, some days (emphasis added).
>
> —Daniyyel/Daniel 12:29–33,
> from the Yerushalayim Tanakh

I am afraid we are now nearing the end of our age, and the Christian church has fared no better than Israel. We have failed to be the light on the hill. Many of our churches no longer believe that the Word of God is the literal, inspired message of our creator. They also have succumbed to the spirit of this age and changed the truth into a moneymaking racket for the greedy and corrupt leaders. The Roman Catholic Church is reeling with innumerable cases of their priests raping children. I need not elaborate; it has become an international scandal.

Were Jesus to walk our streets today, He would be overturning the tables of the moneychangers in Christianity and not just those who peddle their Christianized pabulum on television and live in palaces, but also those who are more concerned with building their own kingdoms rather than the kingdom of God. It is the inevitable march of corruption intrinsic to our human history. This is the fundamental reason for a decentralized and limited government, even in the church.

But this age will soon be over, and the Davidic Kingdom will establish righteousness on all the earth. In that day, the smokescreens used by the Enemy of Man to deceive humankind will end. No longer will the merchants of the earth rule through oligarchies. No longer will monopolies extort the working person. No longer will despots rise to power and oppress their people. No longer will the working person be enslaved and robbed by the elite in power. No longer will nations war against one another in order to greedily annex more wealth and power. Jehovah Tsidkenu will rule with righteousness and justice. In that glorious day, Jerusalem will be known as the center of the world, the city of truth, where the only righteous power can justly rule in a centralized form of government. He alone is the sinless one, the righteous right arm of God.

Our National Responsibility

Our responsibility as Christians and Jews is to effect change toward righteousness and justice in each of the nations in which God in His providence has placed us. Therefore, our role is to help nations fulfill their God-given responsibility to protect the God-given, basic rights of each of those citizens.

We must have the courage to stand against those who oppress our brothers and sisters in every part of the world. We cannot turn a blind eye to a person who is being oppressed without knowing that it will come back to haunt us some day. Abraham, Isaac, Jacob, and Moses taught us that we are our brother's keeper!

As a nation and as individuals, we have a global responsibility to oppose tyranny and the heartless abuse of our brothers and sisters everywhere by any moral means possible. If we continue to show economic partiality to the wicked rulers of tyrannical nations at the expense of the subjugation and suffering of their oppressed and afflicted citizens, God will surely judge us.

God takes His stand in His own congregation;
He judges in the midst of the rulers.
How long will you judge unjustly
And show partiality to the wicked?
Vindicate the weak and fatherless;
Do justice to the afflicted and destitute.
Rescue the weak and needy;
Deliver them out of the hand of the wicked (emphasis added).

—Ps. 82:1–4

There are two major reasons for this travesty. The first is greed. And the first is the reason for the second—an underlying globalist agenda that seeks to align the nations of the world in a single financial system that will facilitate the establishment of a single political system.

Christians and Jews must be wary of anyone who espouses a globalist ideology that rebels against the order ordained by God since the Tower of Babel. The subjugation of our Constitution to a higher global government is, indeed, a dangerous gamble. It will not effect the peace they promise through their globalist propaganda, whether it is from a supercapitalist agenda or a collectivist agenda.

At the same time, we must be careful not to revert to the isolationist notion that leads to a selfish and narcissistic form of provincialism. God has called us to be separate in our behavior and actions but not to be isolationists. We should advocate global consciousness without destroying national distinctiveness.

We are to be *in* the world but not *of* the world. Isolationism is a morally indefensible posture in light of our common heritage, for all humankind springs forth from the same parents at least 10 times. All of us have in common the very same forefathers, from Adam to Noah. From a physical perspective, we are all one family.

Our Judeo-Christian worldview demands of us a much higher regard for true global consciousness than that of the globalists with a naturalistic presupposition. We are our brother's keeper and therefore responsible for the welfare of all humanity. Strangers are, after all, just family we have yet to get to know.

We are to be the champions of justice, equality, and freedom for our entire human family. These are God-given rights to every single member of humankind, and our human governments have been charged with the specific task of securing them and administering justice.

For this reason, no nation is an island unto itself. We are members of a greater family and ought to establish proper relationships with these nations, providing assistance when it is requested and ensuring mutual defense. But our national distinctiveness should not be sacrificed. This is nothing more than the logical extension of our relationship as individuals to one another. We should not sacrifice our individual distinctiveness, as championed by the collectivist dogma, but rather establish unity while securing and protecting our individual unity in the midst of diversity.

We must not fail to understand that in order to accomplish this, power should be kept as decentralized as possible in order to maintain and secure our God-given rights and freedoms. The further the seat of power rests from citizens, the more apt leaders are to abuse their power and the less able we are to prevent it or resist it.

There is no foolproof system of government because in every system there is the human component. And as a result of our fallen nature, power inevitably corrupts any human system. Every great nation throughout every generation has gone through a similar process from its birth to its golden age, and then to its deterioration and demise.

We can observe this in the history of the Israelites as they struggled from slavery and bondage in Egypt to crossing the River Jordan and becoming the nation of Israel. And then we can observe their disintegration until they were dispossessed of their land and disbanded as a nation. But that also applies to every Gentile empire that has risen and fallen, without exception.

It has been noted by others that there seems to be a pattern that is repeated over and over in every human civilization as they move from slavery to liberty and then back to slavery. Here is the simple progression of this pattern:

1. The period of slavery or bondage
2. The period of the awakening to the need to take action
3. The period of the attainment of faith in the justice of the fight
4. The period of the attainment of the courage to battle for liberty
5. The period of the joy of the attainment of liberty
6. The period of thankfulness for the blessing
7. The period of abundance
8. The period of complacency
9. The period of apathy
10. The period of decadence
11. The period of lawlessness
12. The period of tyranny and the return to slavery

This historical process is not cyclical as some have stated, but rather it is in the form of a wave. Time had a beginning (alpha point) and marches forward toward a conclusion (omega point) at the consummation of the ages.

It is our human nature that fowls every human institution in time. Our forefathers understood this explicitly. And our great nation was designed with every possible check to our human nature that they could think of, not only to ensure as much as humanly possible a just form of government but also to safeguard the future from the rise of tyranny.

For this reason, the American experiment attempted to provide a checks-and-balance system that our Founding Fathers hoped would arrest this inevitable corruption by thwarting the singular control of power through the separation of governmental controls into three separate branches.

These are the legislative, executive, and judicial branches of our government. Even Congress was split into two houses that would provide representation for the wealthy landowners through the Senate and for the working middle class and the poor through the House of Representatives. It was also designed to keep the larger more populous states from ruling over the less populous, smaller states.

These mechanisms have worked to one degree or another, as we will discuss, but what our forefathers could not design was the moral fiber of later generations. The ancient cosmic battle between good and evil rages in every generation and in every nation, without exception. Our choices have consequences, not only for the individual but also for the collective of our culture. Like endless ripples from a stone crashing through the surface of a still pond, our immoral actions influence others and eventually result in lawlessness.

Today, we find our Western culture in the 10th step and rapidly closing in on the 11th. There is no time for vacillation. We must act in a concerted effort to reverse this trend or we will soon find that it will be too late.

For these many reasons, our forefathers sought to keep power decentralized as much as possible. As a matter of fact, most early Americans favored an isolationist policy that tried to keep us from being entangled in the affairs of European nations. Although I certainly understand the wisdom in that, from a Judeo-Christian perspective, we must consider that we are our brother's keeper, and we do have a social responsibility to our human brothers and sisters globally.

Most wars fought by humanity are senseless wars borne from greed. World War I was certainly such a senseless and needless catastrophe. But when madmen bent on global dominance rise to power in our world, those who stand for liberty must rise up in opposition. Such was World War II. There are times when we must defend our liberties and keep those with globalist-colonialist ambitions from realizing their selfish dreams.

But our global responsibility does not entail the dissolution of nations in order to form a global government. We must be careful not to fall victim to the globalist agenda and run off the other side of the cliff. Both extremes are equally disastrous.

We must not allow our greed to overshadow our moral imperative to stand against oppressive forms of government. We must not allow totalitarian and ruthless governments that enslave people to profit from our technology or trade. We must not use our financial power to financially enslave other countries. And we should resist with every diplomatic tool the strengthening of those autocratic and despotic governments, even if it means an economic hardship for us.

Sadly, this has not been our course. Solzhenitsyn warned us of the potential disaster of this myopic policy if our greed for profit overrules our moral rectitude and economic pragmatism becomes our god. Unfortunately, the West for the most part has been blind to this reality; we have viewed foreign policy through materialistic spectacles far too long.

Tyrants and despots have been propped up in Central and South American governments, as well as Africa, the Middle East, and Asia, by American policies that benefited materialistically from this unholy union. For this reason, Castro gained support among Cubans in his revolution against the American puppet Batista. For this reason, Nicaraguans supported the Sandinistas against General Somoza, and we could continue ad infinitum. Solzhenitsyn states in *Warning to the West*,

I must say that Lenin foretold this whole process. Lenin, who spent most of his life in the West and not in Russia, who knew the West much better than Russia, always wrote and said that the western capitalists would do anything to strengthen the economy of the USSR. They will compete with each other to sell us goods cheaper and sell them quicker, so that the Soviets will buy from one rather than from the other. He said: They will bring it themselves without thinking about their future. And, in a difficult moment, at a party meeting in Moscow, he said:

"Comrades, don't panic, when things go very hard for us, we will give a rope to the bourgeoisie, and the bourgeoisie will hang itself."

Then, Karl Radek, whom you may have heard of, who was a very resourceful wit, said: "Vladimir Ilyich, but where are we going to get enough rope to hang the whole bourgeoisie?"

Lenin effortlessly replied, "They'll supply us with it."[2]

It seems that Lenin's prophecy is being fulfilled, not through the communists but rather the radical Muslim nations. We have, indeed, sold them the rope with which to hang us. We have armed them against one another, and now they have turned those weapons on us. And even more alarming is that we continue to arm them with weapons and enrich them through the oil industry, even when they have turned those weapons on us. The bottomless greed of the Western merchants is being remunerated dearly and daily by the shedding of our blood.

Lenin also correctly understood that the corrupted form of super-capitalism was hijacking true capitalism and perverting the principles of free and fair trade. But his solution was equally catastrophic since he did not acknowledge the three fundamental principles adhered to by our Founding Fathers—principles that alone could create a just form of government. Collectivism is not the solution. Nevertheless, Lenin's insight into the corruption and bastardization of capitalism was quite accurate. Monopolization has destroyed the capitalist ideal of free competition. Henry M. Christman states,

The old form of capitalism has had its day. The new form represents a transition towards something. It is hopeless, of course, to seek for "firm principles" and a "concrete aim" for the purpose of "reconciling" monopoly with free competition.

"This starting point could perhaps be placed at an even later date, for it was only the crisis" (of 1900) "that enormously accelerated and intensified the process of concentration of industry and banking, consolidated that process and more than ever transformed the connections with industry into the monopoly of the big banks, and made this connection, taken individually, much closer and more active."

Thus, the beginning of the twentieth century marks the turning point at which the old capitalism gave way to the new, at which the domination of capital in general made way for the domination of finance capital.[3]

The process of the consolidation of wealth into giant monopolies that control banks, power, transportation, health services, the media, and even the food industry has continued unabated, in spite of some futile attempts by various nations to redact antitrust laws. Here in America, for instance, the celebrated antitrust laws under Theodore Roosevelt have proved to be as useless as lips on a chicken.

But in the West in the last half of the twentieth century, it has taken a new twist. Western merchants have realized that the fastest way to fleece the flock is through socialism. With the rise of the Soviet Union and communist China, they witnessed the deluding power of the collectivist ideology. The illusion of the dictatorship of the proletariat was quite compelling to the abused laborer who was being squeezed for every last drop of sweat to make the merchants richer. Millions died in revolutions that only managed to substitute one group of power elite for another. When these countries, filled with hungry peasants, raised up in revolt, the blood of merchants and nobles flowed in the streets.

Here in America, due to the division of power designed by our Founding Fathers and the liberties that allowed a strong middle class to flourish, merchants were leery of grasping control of our government too quickly. After all, our nation is by far the most heavily armed civilian population in the world. It is for that reason that their attack on our Second Amendment has not ceased but instead has picked up momentum lately.

Merchants took their first step in 1914 during Woodrow Wilson's presidency when they took from Congress (the representatives of the American people—we the people) the right to control our nation's finances and established the Federal Reserve. That was a cunning master stroke that brought an oligarchy into power in our government. We can thank Woodrow Wilson's progressivist agenda for that betrayal of our constitutional rights as a people to control the economy.

The wily merchants, after the communist revolutions that spread bloody conflagrations throughout our world, began to see that if people are squeezed too hard, it might backfire. Yet their greed was hard to repress. Their diabolically clever plan was to create a depression that would sink most of their competitors into ruin. When all seemed lost, they would step in and buy those companies at pennies for the dollar. And the consolidation of their financial grip on the nation thus took huge strides.

Many in America were quite disturbed over this monopolization of the American economy, but then the Second World War came and drew our attention away from domestic concerns, rearing its ugly head and threatening the freedom of all in the West. Americans doubled down and joined the effort to stop Hitler, Mussolini, and Japan from continuing their global rampage. Our factories hummed with work as we began to arm England, France, and Russia. Our economy once again began to move, and the oligarchs reaped an enormous windfall from selling war goods.

The middle class was, in essence, saved by the war. But the military industrial complex blossomed in power and wealth. After

World War II ended, the Soviet Union became our greatest threat. The Cold War began, and once again the military industrial complex profited handsomely from the nuclear escalation of the Soviet Bloc and the free world. The growth in technology once again saved the middle class. But as the Cold War ended, the oligarchs realized that there might be a better way to extract more from the citizens. That way was socialism. Their sights expanded to globalism.

They began to see that socialism would be even more profitable for them. They realized that enormous amounts of money can be made by financing national governments that are deep in debt because of their socialistic promises of free stuff. The myopic sheep bellow happily as they are corralled into their pens and sheered to their skins. They are no longer free to roam. Naked shall they be, and the middle class of the entire world will disappear the day the New World Order consolidates and centralizes the economic and political power of the entire world.

At the present time, national governments are the debt collectors for the international financiers of the World Bank. They collect the taxes and then pay the bankers an enormous interest rate on their overblown national debts. But if they had their way, all of this would be significantly simplified by having one international centralized collection government and a world tax for every human being on the planet—a New World Order.

There is, however, a fly in the ointment called Islam. Islamic clerics want to have their own global government. Their religion urges them to conquer every mountain and every valley on earth for Allah. Their religion urges them to force by the edge of the sword every human being to submit to Allah and their Sharia law. The card up their sleeve is that they sit on more than 70 percent of the world's oil reserves. Their militaristic-colonialist religion is, indeed, a powerful weapon that, I fear, is heavily underestimated by the arrogant atheists in the West who trust too much in their technological prowess.

A war is coming, whether we like it or not. It looms ever closer on the horizon as we approach the end of this age. I fear that Turkey is positioning itself to once again reestablish the caliphate in order to unite the Sunni Muslims. The chess pieces are moving alarmingly fast. What is the solution to this problem? I am afraid I am not smart enough to know that answer from a temporal perspective. The only answer I see is the return of the Messiah.

A friend asked me, "If you had a magic wand and you could change the political condition in the world, what would you wish for?" I would wish for the freedom of all nations to compete in the free market without the unfair power of monopolies. I would wish for unity in the midst of diversity. I would wish for true tolerance among all races and people. But I would not wish for the dissolution of nations.

While it is true that we should avoid the divisions and hatred spawned by supernationalism, we should not seek unity at the cost of sacrificing our national differences, which provide for diversity. Unity should be sought in the midst of diversity. Unity does not mean uniformity. It is not monism or consensus omnium. Unity between nations ought to be in the midst of the freedom of diversity.

Nations can unite in purpose and in action, but they should remain sovereign and independent nations. That allows us to keep global power decentralized. To tie our nation to a coalition that requires a higher authority over it is to centralize power and cement the death of the middle class and the death of personal freedom.

History has left the undeniable record for all who care to see it that as power in governments became more centralized, the freedoms of individuals suffered in direct proportion. Even within the independent nations, the centralization of power should be limited to the least common denominator necessary to provide national defense and domestic tranquility. That is what I would wish for.

Sadly, this has not been the case in all human history. Most nations have historically sought to centralize as much power as possible in order to amass the wealth and military prowess necessary to conquer others by the greed of those in power. That has been a great concern for Americans from the very first day of our independence until the Great Depression, which turned everything on its ear. But we are getting ahead of ourselves.

Notes

1. Joseph Stalin, quoted in Robin Cross, *Fallen Eagle: The Last Days of the Third Reich* (London: Thistle Publishing, 1995), 21.
2. Alexander Solzhenitsyn, *Warning to the West* (New York: Ferrar, Straus and Giroux, 1976), 13.
3. Henry M. Christman, ed., *Essential Works of Lenin: "What Is to Be Done" and Other Works* (New York: Dover Publications, 1987), 202–203.

CHAPTER 6

● ● ●

THE CENTRALIZATION OF POWER

Global centralization of power can be attained several ways, some more violent than others. But in the end, they all bring tyranny. It is the consequence of humanity's rebellion toward the judgment of the Tower of Babel when God established the nations to avoid this very dilemma. Any attempt to bring a single nation above all others or to make a cartel of nations to rule over the world is thus a corruption of God's original intent in the establishment of independent nations. It is called globalism.

However, we must also realize that the complete decentralization of power is also a counterfeit of God's ordained system of government. The attempt to do away with all government is anarchism, which leads to chaos and inevitably tyranny. God's purpose for the establishment of national governments is to ensure justice and provide safety for its citizens. Any time in history when there was a power vacuum to maintain peace, villains of all sorts preyed upon the weak.

Our human nature is such that it is impossible to survive peacefully without a restraining police force that can secure the innocent from those who would abuse them. But that power must be kept to the bare essentials in order to keep them from also abusing the citizens. An unrestrained police force can lead to the same heartless abuse of the innocents. Anarchy is the straight road to violence, but an overly powerful police force can also lead to that same road of oppression and violence.

Those who promote the globalist agenda under the mistaken idea that it will bring peace to the nations are, in fact, setting up a system that can be easily manipulated by the financial elite to do the exact opposite. It will lead to the tyrannical control of all nations and citizens. It will lead to the financial exploitation of the masses. Such concentrated power cannot be safely wielded by any human or groups of humans without greed infecting and controlling the outcome. To think otherwise is utterly naïve. It is to ignore the entire historical record of humanity that, without exception, has shown that we are broken human beings incapable of resisting the temptations that power provides. It is pure folly.

We must resist the globalist plan to blur humanity into a single collective organ, devoid of individual uniqueness and value, in order to economically and politically exploit us. It must also be noted that collectivism in practice has historically always inevitably led to the centralization of power and thus equally resulted in tyranny. It is an indisputable axiom that the more power that is given to either a person or an elite group of people, the more corrupt that person or that group will become. For a people to be free, government must be restricted and limited in power.

Most of us need not be convinced that absolute power corrupts absolutely. It is for this reason that God separated humanity into nations during the judgment of the Tower of Babel by confounding the languages and fragmenting power. Ever since, Satan and his minions have been working feverishly to reunite humanity in order to gain control of our planetary government. Make no mistake; the real power behind any globalist agenda comes from the dark angels.

Their plan is no secret. It is out in the open for those who have eyes to see. Since that time, the call of the occult, mirrored in the Masonic motto, has been to gather that which was scattered in order to consolidate knowledge and power and realize their plan on earth. Historically, none can argue that in the last century, collectivism has been their most efficient ideology and has invariably led to the total centralization of power and the diminution of the intrinsic rights of the individual.

There is not a single instance where communism has not led to this unfortunate consequence, which, as we will discuss later, also leads to economic failure. And for that reason, it has been one of the primary venues for the globalist agenda of the demonic hierarchy. They have adapted their tactics as they have redressed the same old lie to become more appealing to the modern masses. Collectivism has been reshaped into a more palatable form called socialism and progressivism. But socialism and progressivism at their foundation are, in essence, equally collectivist and lead to the centralization of power and eventual tyranny.

Any attempt to remove power from the local sphere and take it to a centralized point is a move toward tyranny and away from personal liberty. We must never forget that the further removed from the general public our rulers become, the more likely they are to be self-serving and tyrannical. It is the natural by-product of our human condition due to the fall of humanity.

For this very reason, God pronounced judgment at the Tower of Babel in order to safeguard our liberties by decentralizing power and forestalling tyranny. Both supernationalism (whether in the form of the Nazi Party, the Islamic movement, or an oligarchy) and collectivism (whether in the form of communism, socialism, or progressivism) attempt to override this mandate instituted by the creator shortly after the Great Flood. They inevitably bring upon humanity the destructive consequences of the centralization of power in their globalist agendas.

Today, both socialism-progressivism and the colonialist-globalist ideology of radical Islam are threatening the freedom we have in our Western culture. History has demonstrated with crystal clarity that the interests of bureaucrats in distant capitals, far from individual citizens, will always move toward greater repression and less personal liberty. It is utterly naïve to think that the socialist ideology of appropriating or completely controlling people with burdensome government regulations for all private businesses will lead to nothing less than a monopoly with the added repressive components of an army and their secret police to brutally enforce their exploitation of the masses.

Our cherished concept of self-government is at stake. This paramount ideal for our nation is a principle for which many have suffered and died. We the people give consent to those who rule over us, as long as they maintain their rule within the proscribed limits and boundaries imposed by our Constitution.

Those limits were designed to fractionalize and decentralize authority in order to forestall the accumulation of too much power into one or even a few hands. The ideal of placing the lion's share of power with the local government and not the federal government was specifically designed to protect the general citizenry from an unscrupulous elite wanting to monopolize our national affairs.

Moreover, in this form of government designed by our Founding Fathers, leaders are responsible for maintaining, protecting, and executing the absolute standards proscribed in our Constitution. That is, the law is supreme, and those who govern are subject to the same laws as the governed. And they are to be held accountable to us—the governed. For that reason, the House of Representatives was given oversight over the executive branch. Sadly, the political elite today believe they are no longer under the laws that apply to the rest of the nation.

The executive branch under Barack Obama used the IRS as a political tool to attack political adversaries. Many elected officials, including Hillary Clinton as Secretary of State, consistently broke laws, and the Department of Justice and the FBI refused to punish the guilty. The power of the federal government is now quite despotic. Lady Justice, depicted in the statue outside the U.S. Supreme Court building, is holding the balance of justice and has a scarf around her eyes, but she is no longer blind to those she judges. The scarf should be removed from her eyes, and her hands should be tied behind her back with that same scarf. She is now the tool of the elite to protect their minions and destroy their enemies.

But in order for our government to function as our forefathers intended, most of the power of government should reside in the local community. Central government should have only as much power as absolutely necessary for it to provide the common defense of the states. That includes protecting our borders, regulating immigration, promoting the overall welfare of citizens, and providing for the domestic tranquility of the nation.

The more power we give to the central government, the less power the common person has to influence the direction of government in any practical way. For that reason, our Founding Fathers created the Articles of Confederation ratified in 1781, which kept governmental power completely decentralized. So much did they mistrust the potential danger of the power of a federal government that they purposefully made it infinitely weak. Unfortunately, it was too decentralized, and later the Articles of Confederation were replaced by the Constitution because there was not enough power for the federal government to accomplish its duties without the power to garner taxes.

If our Founding Fathers could have seen the overblown bureaucracy and entangling tentacles of power into which the federal government has now morphed, they would be starting a new revolution. At the beginning of our nation, taxes were levied on the consumption of products. If a person had greater wherewithal to purchase more things, then that person paid more taxes. There was no income tax.

People were not taxed for what they earned but for what they spent. We have moved far from this foundational principle as our superfederalists have blinded our nation from these basic truths known to all citizens at the time of the founding of our nation.

The War Revenue Tax of 1917, levied upon the American people during World War I, was supposed to be repealed after the end of the war. Instead, our federal government has steadily increased the percentage of our income tax to meet the burgeoning federal bureaucracy that has grown as a cancerous tumor that will eventually kill our liberties.

At the end of World War II, the rising danger of the Soviet Union and the reality of the Cold War's military costs served to justify the continuation of the tax. Ever since, our nation has been moving further and further toward superfederalism, which has served to insulate our central government from the common person.

Make no mistake, the danger to our global peace has not disappeared with the dissolution of the Soviet Bloc. Russia is once again amassing nuclear weapons and perfecting its delivery systems. It has hypersonic intercontinental missiles with nuclear warheads that would be quite difficult to defend against. Russia's President Putin is attempting to reconquer the areas that were lost when the Soviet Bloc fell apart. His involvement in Syria backing Riad al-Asaad and his backing of the Iranian government and the Maduro socialist regime in Venezuela are a huge danger to the West.

China, after reforming its collectivist ideology and accepting some capitalist ideas, is growing in strength and power as no other nation in the world. It has one-fifth of the world's population, and its army is extremely well equipped with nuclear weapons. China is now artificially building islands in the middle of the Pacific Ocean's shipping lanes in order to create fortresses in the middle of the international shipping lanes of the Pacific Ocean. That is quite a dangerous development that our nation has yet to address. In the meantime, we have a $500 billion yearly trade deficit with China, which is attempting to become the richest and most powerful nation in the world.

Today, the radical jihadist globalist mentality of Islam has become our new great nemesis. In the shadow of the Islamic threat from the Middle East, the collectivist ideology is rapidly gaining territory in the West. The Soviet threat has not disappeared. Russia is on the march again and has retaken Crimea. As of August 2016, it was quietly building its forces on the border of Ukraine. I fear for the future liberty of the people of Ukraine.

North Korea, a proxy to the Chinese state, has embarked on a nuclear race and the building of intercontinental missiles that could reach the United States. The unpredictability of North Korea's despotic rulers may end up being the spark that begins another global war.

Incredibly, Obama gave Iran a free pass for the construction of nuclear bombs with the signing of the disastrous Iran nuclear deal. A nuclear bomb in the hands of Iran, which exports the most terrorism in the entire world, is crazy foolish to say the least. Obama also handed back some $150 billion of frozen assets that will most assuredly be used to foment terrorism against the West. Iran is attempting to create a land bridge between Iran and the northern coast of Israel; it goes through Iraq, Syria, and Lebanon. All of these global threats are very real. But our domestic threat is no less lethal as the popularity of socialism increases among our youth.

The Day Americans Were Betrayed

In America, the single most important step away from our cherished ideals of self-rule came at the turn of the twentieth century when the right of the people to oversee their monetary system was given to international bankers. The watershed moment for the transformation of this process came when Woodrow Wilson signed over our financial rights—rights our Constitution had reserved for the representatives of the people—to a private consortium of banks. The establishment of the Federal Reserve, which is not a governmental agency but a private banking trust, was the milestone in our nation's betrayal. This one single act of treason has given the financial elite the real power behind our government. It was their foot in the door to total power in America.

Clever are the ploys of these robber baron merchants. It was the artificially created financial crash in the first decade of the twentieth century that strong-armed Woodrow Wilson to sell out our nation to the bankers. To be fair to him, he later came to understand the true nature of what he had done. In a letter, he admitted that he was hoodwinked into thinking it would give the nation financial stability. It did nothing of the sort. It put the wolves in charge of the henhouse. That weighed heavily upon his shoulders. Wilson died of a massive stroke before his term was over. America was hoodwinked, and we the people no longer have control of the monetary system as our Founding Fathers designed.

More and more, politicians became indebted to the bankers who ruled the roost. Their influence caused the government to favor certain companies and unfairly discriminate against other competitors. Instead of the government protecting the rights of all individual entrepreneurs to prosper, carte blanche was given to a few who unjustly undercut their competitors. The birth of the giant trust companies spelled the beginning of the end for the mom-and-pop store, the family farm, or the local bank.

Later on, President Franklin D. Roosevelt put the final nail in the coffin when he instituted measures that undercut much of what our Founding Fathers had died to secure. A second manufactured economic crash brought about the Great Depression. Unfortunately, Wall Street industrialists and financiers took advantage of this economic instability to strike down what little antitrust measures were previously redacted.

Instead of keeping the federal government at a minimum to fulfill its responsibility, the federal government was expanded to a previously unfathomable size. Consequently, the tax burden on the common person increased proportionately. The desperation in our nation created by the Great Depression provided the leverage necessary to cause the public to turn a blind eye to these drastic draconian measures as they frantically sought solutions to the grave economic hardships they faced.

In the meantime, the government began to swell, increasing the workforce but also increasing the tax burden on the working person. For all intents and purposes, the measures created to prevent the burgeoning of our national debt and the creation of giant monopolies through antitrust laws have since been effectively neutralized or sidestepped. At this point in our history, capitalism morphed into supercapitalism, and both our political parties have been controlled by them behind the scenes ever since.

During the last two decades of the twentieth century, we saw the consolidation of the banking industry to a powerful few. Hundreds of savings and loans faltered and were scooped up by the big banks. Then, in the first decade of the twenty-first century, we witnessed the third artificially manufactured economic crash of 2008. The noose around the neck of the middle class was cinched ever tighter. The idea that our government would take our tax dollars from the working person to bail out huge financial institutions that, out of sheer greed, brought them to the edge of bankruptcy is a travesty I will never understand. Why should we bail out those who financially rape us?

We will cover these three economic crashes more thoroughly later on. My purpose in this section is to give the reader a bird's eye view of the events that are bringing our nation away from our founding precepts and into alignment with a globalist agenda.

I foresee a time in the not too distant future when this same tactic will be used to further promote globalist aims. Perhaps the rising cost of fuel created by a Middle East crisis with a nuclear Iran will become the catalyst for their next step in the diminution of our nation's middle class in order to enable them to dissolve us into the globalist agenda. Or perhaps they will manufacture another crash in the stock market. Only time will tell. Sadly, most of us will capitulate to anything that promises physical prosperity in a time of crisis.

Prior to the twentieth century, the Jeffersonian notion of the decentralization of power had been cherished as an immutable American political doctrine, deeply rooted in the conscience of American culture, from the major metropolises to the Heartland—from shore to shore. It had been one of America's most powerful sentiments, but the public education system has brainwashed our children, who typically do not even know this. Jefferson was heavily indebted to British banks and was quite afraid of the banking industry taking control of our nation. He wrote extensively of this potential danger, and almost every American regarded these Jeffersonian political views as gospel before the progressivist indoctrination system took over our educational system and slowly rewrote American history. David M. Kennedy explains in *The American People in the Great Depression*:

> Many of them still clung to the political faith of their fathers, to simple Jeffersonian maxims about states' rights and the least possible federal government. They reverenced a balanced budget as the holiest of civic dogmas.[1]

But, even though we have come a very long way toward federalism, at least most of us in the past have been able to rely on the overarching protection of our Constitution and the freedoms our Bill of Rights so elegantly declares. But they may not last very long, either. Those who promote their own particular version of the globalist agenda are also attacking us viciously in a classical pincer movement from two sides to destroy the separation of powers.

The Global Charter and the Reinterpretation of the Constitution

Peace on earth has been people's most cherished dream and at the same time the most elusive to attain. Perhaps the most sinister lie being promoted by the architects of globalism is their supposed reason for disarming the nations and consolidating military power in one centralized world government in order to attain global peace. This call to disarmament, of course, always begins first with disarming the people. Only then, they claim, can humanity be free of war.

Many well-meaning idealists are drawn to this argument because of their disdain for the horrors of wars. I certainly share in that disdain for the brutality of war and terrorism. As global conflicts and the violence of terrorism increase in our world, humankind will be pressured and deluded into thinking that this will be the only pragmatic solution that will finally grant them personal safety and at long last bring peace to our world.

There is, however, a glaring problem with this myopic illusion. It fails miserably to account for the entire history of humanity and for the fundamental nature of humans, who are prone to abuse power for selfish interests. When such awesome power is centralized in a few, the vast majority of the common people have always, in every historical instance, been oppressed and victimized.

Without exception, every conqueror has offered a modicum of personal peace from bandits and lawless people to the vanquished, but at the deep cost of financial slavery and the loss of personal freedoms. We can point to the so-called Roman peace offered by the Roman Empire as a prime example of this historical reality. The Romans' recipe for keeping the masses in control was called bread and entertainment. Truth be told, it worked. As long as people had a modicum to eat and their minds were occupied with the year-round entertainment provided in the coliseum, their minds were basically neutralized. It is no different today. The elite in power always seek to wrest more and more power at the cost of people's personal liberties.

It is for this reason that the Judeo-Christian worldview offers humankind the only true understanding of their human frailty as fallen creatures in a broken world. Because most modern people, hoodwinked by the relativistic ideology, reject this fundamental understanding of the fallen nature of humans central to the Judeo-

241

Christian worldview, humanists and socialists are incapable of providing a fundamental worldview to build a government that can realistically protect us from ourselves. That was one of Lenin's gravest mistakes. The only thing collectivism accomplished in Russia was the enablement of the ruthless to exterminate all whom they believed could ever be a threat to their power. The Soviet purges of Stalin with their stark oppression and abuse of basic human rights through communism stand in stark contrast to the American Revolution and the freedoms we gained.

It was the Judeo-Christian background of America's Founding Fathers that spurred them to provide our citizenry with the protection of the Second Amendment to the Constitution that assured them the right to bear arms. The purpose for an armed citizenry was considered indispensable as a deterrent to the abuse of power by the federal government—an inevitable probability for those who understood the historical reality of fallen humanity. The right to bear arms was not meant to be a right to hunt for food or sport but rather a right to defend their hard-earned liberties from those who would abuse the power allotted to them by the people. It was meant to protect them from their own leaders. It was meant as a check and balance of the federal government.

In direct opposition to the politically correct propaganda emanating from Washington, DC, is the belief that the greatest danger to world peace and individual freedom for all humanity is, has been, and will be the centralization of power. The blood of the victims of those tyrants who have been demonically inspired to conquer the world has stained every epoch of our human history. Nothing has changed the demonic illusion that has existed since the time of Adam.

Ever since the judgment of the Tower of Babel when God in His wisdom divided humanity into separate and distinct nations to avoid the great evil of the centralization of power, Satan has been hard at work to unite humankind again under a single government.

Countless conquerors have been aided by the dark angels to attempt a global union of all nations. So far, none have been able to manage the complete control of the entire world. But our prophets have foretold that with no uncertain terms at the end of this age, one will conquer the whole world.

Forces from him will arise, desecrate the sanctuary fortress, and do away with the regular sacrifice. And they will set up the abomination of desolation. By smooth words he will turn to godlessness those who act wickedly toward the covenant, but the people who know their God will display strength and take action. Those who have insight among the people will give understanding to the many; yet they will fall by sword and by flame, by captivity and by plunder for many days. Now when they fall they will be granted a little help, and many will join with them in hypocrisy. Some of those who have insight will fall, in order to refine, purge and make them pure until the end time; because it is still to come at the appointed time.

Then the king will do as he pleases, and he will exalt and magnify himself above every god and will speak monstrous things against the God of gods; and he will prosper until the indignation is finished, for that which is decreed will be done. He will show no regard for the gods of his fathers or for the desire of women, nor will he show regard for any other god; for he will magnify himself above them all. But instead he will honor a god of fortresses, a god whom his fathers did not know; he will honor him with gold, silver, costly stones and treasures. He will take action against the strongest of fortresses with the help of a foreign god; he will give great honor to those who acknowledge him and will cause them to rule over the many, and will parcel out land for a price.

—Dan. 11:31–39

"Then I desired to know the exact meaning of the fourth beast, which was different from all the others, exceedingly dreadful, with its teeth of iron and its claws of bronze, and which devoured, crushed and trampled down the remainder with its feet, and the meaning of the ten horns that were on its head and the other horn which came up, and before which three of them fell, namely, that horn which had eyes and a mouth uttering great boasts and which was larger in appearance than its associates. I kept looking, and that horn was waging war with the saints and overpowering them until the Ancient of Days came and judgment was passed in favor of the saints of the Highest One, and the time arrived when the saints took possession of the kingdom.

"Thus he said: 'The fourth beast will be a fourth kingdom on the earth, which **will be different from all the other kingdoms and will devour the whole earth and tread it down and crush it.** *As for the ten horns, out of this kingdom ten kings will arise; and another will arise after them, and he will be different from the previous ones and will subdue three kings. He will speak out against the Most High and wear down the saints of the Highest One, and he will intend to make alterations in times and in law; and they will be given into his hand for a time, times, and half a time. But the court will sit for judgment, and his dominion will be taken away, annihilated and destroyed forever. Then the sovereignty, the dominion and the greatness of all the kingdoms under the whole heaven will be given to the people of the saints of the Highest One; His kingdom will be an everlasting kingdom, and all the dominions will serve and obey Him'"* (emphasis added).

—Dan. 7:19–27

This King of the North will tread the whole earth for the first time in human history, but it will only last three and a half years until our Champion comes to claim His rightful throne in Jerusalem. The irony is that in the middle of the Great Tribulation of seven years (known as Jacob's Trouble in the Hebrew Scriptures), this usurper will announce his global victory when he conquers Jerusalem at the same time that God says the Day of the Lord will begin. And for the next three and a half years, God's wrath will cleanse the earth of wickedness while God provides a place of hiding for Jewish believers underneath the Eagle's wings. From the ashes of the death of the second earth will the third earth rise, and Israel will become the priests of God once again in the Davidic Kingdom.

> *Then another sign appeared in heaven: and behold, a great red dragon having seven heads and ten horns, and on his heads were seven diadems. And his tail swept away a third of the stars of heaven and threw them to the earth. And the dragon stood before the woman who was about to give birth, so that when she gave birth he might devour her child. And she gave birth to a son, a male child, who is to rule all the nations with a rod of iron; and her child was caught up to God and to His throne. Then the woman fled into the wilderness where she had a place prepared by God, so that there she would be nourished for one thousand two hundred and sixty days.*
> —Rev. 12:3–6

> *And when the dragon saw that he was thrown down to the earth, he persecuted the woman who gave birth to the male child. But the two wings of the great eagle were given to the woman, so that she could fly into the wilderness to her place, where she was nourished for a time and times and half a time, from the presence of the serpent.*
> —Rev. 12:13–14

The ageless obsession of greed in people obsessed with a diabolic addiction for power and wealth has brought an unfathomable and tragic cost in blood and suffering to every human generation since the dawn of humanity. It is the stark reality of our fallen human condition in a broken world. The illusion of our modern civilization that science could lead us to an age of peace and prosperity has been made naked by the true impact of an ever-increasing power to destroy not only life but everything else around us.

Less than a century ago for a few dreadful years, our world teetered on the edge of a knife, and Satan's global dominion was almost achieved. Were it not for the arrogance of Hitler to attack Russia before annihilating the Western threat, Satan would have achieved his goal. But that threat comes not just from people with military methods. Every attempt at globalization by the West in the last century, first with the League of Nations and then with the United Nations and the establishment of the world banking system, has worked to culminate in a one-world government with a global currency. Globalization will ever be a step toward an absolute tyranny.

But the power now is exponentially more dangerous than ever before. Never before in human history has humankind reached a level of such sophistication in technology that the power possessed by people in power could wreak the level of carnage and brutality they possess today. I am not speaking only of the destructive capability of modern weapons but of the capacity to completely and utterly control the masses in every aspect of their lives. The ability to spy on all human beings in the world in their very homes has never existed before in our entire human history. It does today.

The danger to our lives and freedoms does not come just from nuclear and biological weapons and terrorists but also by the unimaginable capacity of supercomputers to analyze every bit of information or mega data that governments are collecting. Our modern governments will soon know every detail of every human being they govern in the entire world—what they like or dislike, what they read, what they buy, what they think. These governments will be able through numerous technological gadgets to listen to and watch every private conversation that takes place in the supposed privacy and safety of people's homes. The digital age has abruptly ended our right to privacy.

Such awesome power can be potentially used for great good or for great evil, as is true of all technology. We are entering an age of uncharted waters where a tyrant can have complete and absolute tyrannical control of all humans on the planet. Now more than ever before, the centralization of power has reached an apex of lethality and danger to the personal liberties we so cherish as human beings who were created in the image of God.

If we are to remain free as a people, we must now more than ever before be circumspect in electing our leaders. We must not be deceived by political propaganda and empty bluster. We must know that those who lead us share our fundamental worldview regarding the fallenness of humankind and the need for the checks and balances wisely created by our Founding Fathers in our Constitution in order to hamper the rise of tyranny in our nation.

We must safeguard our personal liberties and maintain the wisely designed division of government devised by our Founding Fathers or we will lose the precious liberties for which so many have sacrificed their lives. We must not allow our politicians to reinterpret our Constitution through Hegelian spectacles, or our great nation will cease to exist as a beacon of liberty and freedom. And if we hope to forestall the fiscal destruction of our economy, we must once again force our politicians to balance the national budget.

The Pincer Move

Today, the globalist threat to humankind is attacking on two fronts. One comes from the West, and the other comes from the Middle East. From the West, the attack comes in these three forms:

    1. The subjugation of our Constitution to a global charter.

2. The tactic of mining our judicial system with progressives and socialists who would reinterpret the Constitution through their pink ideological spectacles.

3. Expanding the parameters created to contain the executive branch by weaponizing our intelligence community against citizens (the Soviet Union did this quite efficiently with the KGB, and Hitler did likewise with his elite SS guards; Cuba also did it with their G-2, and in like manner, so has every other tyrannical government around the world).

All three tactics are equally lethal to our freedoms as intended by our Founding Fathers. The progressivist and socialist elements would have us place our Constitution under the overriding control of a global charter that would supersede any of the principles expounded in our Constitution. In effect, all the freedoms our forefathers bled to secure would be done away with in a single stroke, as the global charter would have preeminence over any national charter. In one fell swoop, everything can change.

Their plan is to reach this global charter through the following three smaller steps that will be thrust upon us as a necessary policy in order to compete in the international market:

1. The weakening of our economy and our manufacturing potential through lopsided trade deals that take an enormous number of industries out of our nation for cheaper labor in foreign countries, which will have a dire effect on our middle class. By reducing the financial power of the middle class, the elites are able to control the mass media and the electoral process.

2. Establishing unions of nations in other parts of the world that create a larger economic bloc and a greater capacity for them to compete with us in the global market. The establishment of the European Union is the first step in this dark highway of the centralization conspiracy. Soon, their economic and

military prowess will dwarf that of almost all other single nations. Nations will be forced to unite in similar unions in order to compete economically and militarily. That battle is currently raging in England as the people wisely voted to get out of the European Union through the so-called Brexit. But the globalists are doggedly stalling and trying to undo the will of the people.

The moment the United States joins one of these unions, our Constitution will be placed under the higher authority of this union charter. With one single stroke, all of our cherished protections guaranteed by our Bill of Rights will be null and void under the authority of the higher charter of the new union.

3. The appropriation of the common grounds of all nations; namely, all waterways and our atmosphere. In the name of ecology, these will be controlled by the United Nations. We already saw this in America as the federal government under Obama sought to gain all the water rights in our nation. Their eventual goal is to establish a global system that will control all water, including lakes, rivers, streams, and seas, through trumped-up environmental rationalizations. We have already seen this process begin in America as the federal government attempted to strip water rights from ranchers.

During the Obama administration, the federal government began to encroach on the water rights of ranchers. E. Wayne Hage, a Nevada rancher, had for decades owned historical rights to water his livestock, irrigate his ranch, and consume it for domestic purposes. Hage fought back and filed a lawsuit in the Court of Federal Claims, maintaining that the U.S. Forest Service and the Bureau of Land Management had infringed on his rights to access the streams and ditches which his water flowed to.

The Court ruled that Hage had vested rights and that the federal agencies had wrongfully deprived him of his rights. The federal agencies then filed at the Federal Circuit Court of Appeals, which subsequently partially reversed the earlier court decision. While the land case was pending in the Court of Federal Claims, the federal government filed a second lawsuit in Nevada with the hopes of getting a better outcome. They argued that the federal government owned the land and the water rights and subsequently sued for damages by Hage's alleged trespassing on their water.

Hage claimed that it was the federal government who was infringing upon his constitutional rights by purposefully denying him access to his water and refusing to process his applications to renew his grazing permits. In a 104-page ruling, the Nevada District Court issued a scathing verdict detailing the federal agencies' "vindictive," "shocking," and "nonsensical" actions against Hage's renewal application for grazing rights. The agencies had blocked Hage's cattle from grazing and watering for over an entire decade.

The grudge against the Hage family resulted in the federal government soliciting and granting temporary grazing rights to other ranchers in the area, where the Hage family had historically ranged their cattle and had priority interest. They went as far as sending notices to people who leased or sold cattle to the Hages in an attempt to pressure them not to do business with the Hage family and threatened punishment of any who would give testimony for them in the lawsuit.

This audacious conspiracy was documented in a March 29, 2013, article by Brian Hodges:

> The feds' behavior toward the Hage family shocked the
> Nevada court, which held certain government officials in
> contempt and found that they had "entered into a literal,
> intentional conspiracy to deprive the Hages not only of
> their permits but also of their vested water rights." The

250

court determined that the agencies' actions "arbitrarily and vindictively" stripped the Hage family of their water rights and statutory privileges and violated their due process rights.

Aside from the government's stunning behavior, the court's determination of Hage's property rights is also significant. The crux of the feds' complaint was that Hage trespassed on federal lands when his cattle grazed on the banks of streams and ditches in which he held stockwater rights. The feds argued that grazing exceeded Hage's right to access and use water running over federal lands, and therefore constituted a trespass. The court disagreed, concluding that ranchers like Hage hold a "forage right" associated with vested stockwaters. That right allows ranchers to drive cattle to water sources without fearing liability for trespass if a cow stops to eat. In short, the forage right recognizes that cattle can wander a bit and engage in some incidental munching while having a sip of water.[2]

The audacious usurpation by the federal government of state-owned property and the natural water rights of those with private property has already led to several standoffs between ranchers and the federal government. Unfortunately, they ended with lethal force being used on an American citizen. This battle has only just begun.

In addition to this very real threat looming ever more ominously on the horizon, the second front of the attack comes from those who have sought to water down the very meaning of our Constitution through the appointment of judges and justices who reinterpret the Constitution through the grid of the relativistic Hegelian-naturalistic paradigm. They are creating a new socialist-inclined order through unnaturally extrapolated interpretations completely at odds with the original intent of the framers of the Constitution.

They have slowly but surely eroded the idea that the elements of the Constitution are to be respected as absolute in nature. They have effectively moved our nation, inch-by-inch, toward the relativistic ideology of collectivism or socialism.

They have attacked our Constitution as an antiquated, 200-year-old anachronistic document that must be modernized in order to meet the needs of our modern world. Their familiar smokescreens are cleverly employed to mask the real sinister intentions behind their propaganda. Using such clever slogans as "the Constitution is a living document," they seek to enable judges and justices to create new laws through fiat interpretations that override the true intent of the framers of our Constitution. For all intents and purposes, they are promoting the idea that all elements of the Constitution are open to reinterpretation through their Hegelian pink magic spectacles.

These progressive reformers claim that the framers of the Constitution did not intend to be intransigent in the particulars of their document. These relativists cleverly promote their propaganda that the Constitution is an antiquated document that needs to be adapted to modern conditions throughout our public educational system and the mass media. Hence, they claim to remain in the spirit of the framers by extrapolating to meet the new demands brought on by a new era.

Clever are the lies of the enemy! But the Spirit of God freely gives wisdom to all who ask for it. Our Constitution is not an anachronistic document that has outlived its usefulness. On the contrary, its paramount principles are as important as they have ever been. And in fact, it can be argued that no other generation since the founding of our nation is in greater need of these liberties due to the rise in surveillance technology that could usher in a monstrous tyranny that would utterly control our every move.

It can also be argued that it is more important than ever since the influence of one nation on another is far more radical; our world has shrunk due to our technological advancements. Therefore, the principles outlined by our Constitution, which has secured the greatest possible freedom ever achieved by a human government in the history of humankind with the exception of ancient Israel, have global implications for those seeking the same freedoms elsewhere.

Open your eyes, my friend, and listen to the alternatives they propose. Understand the seriousness of this matter before the freedoms you enjoy are taken from you. Seek the truth cleverly buried below the highly polished surface of their cleverly crafted smokescreens. It is for this reason that our nation should begin to realize the overarching importance of having strict constructionist judges and justices at every level of the judiciary, ones who will not try to override the Bill of Rights, the most excellent achievement of any nation in the history of humanity.

The Threat of Supernationalism

As stated earlier, supernationalism is a corrupt and perverted form of healthy and balanced nationalism that promotes the false idea that one nation is intrinsically superior to another. It is the fruit of arrogance and pride fanned by Satan in order to foment and justify militarism, imperialism, war, and colonialist aggression against the weaker nations. While it is true that supernationalism has ever been the springboard used to manipulate the masses by those who have fomented wars, yet the dissolution of a true, healthy, and balanced nationalism among the nations would be just as destructive.

The overt agenda of the socialist promoters of globalization has for some time concentrated on promoting the idea that nationalism is the cause of wars. But that is simply untrue and deceitfully misleading.

Nationalism has been branded as a divisive and anachronistic ideology that foments division, hatred, and war. That idea is promoted in order to bolster the goal of establishing an overarching global government that will rob the public of the ability to have any input into the higher echelons of power within a gargantuan and unreachable centralized socialist government ruled by the financial elite.

The danger to global peace does not come from nationalism but from supernationalism and collectivism. The supernationalist nation profits at the expense of other nations. It seeks to conquer and colonize nations it deems inferior. In doing so, global diversity is sacrificed under the military iron fist of the conqueror who imposes a monolithic idea of culture over the vanquished nations.

We narrowly escaped the clutches of supernationalism in the past century when the Allied Forces defeated the Axis Powers. But today, an even more ominous strain is quickly rising to power on our planet. I am speaking of Islam and its intrinsic globalist mandate found in the very text of its holy book, the Koran.

Ravi Zacharias, a prominent evangelical leader born in India, knows this very well. In India alone, 81 million Hindus were killed by Muslims in their attempted genocide from 1000–1500 AD. Christians are duped into thinking that Allah is the same as the God of the scriptures, but he is not.

> Islam is a religion that is academically bankrupt, for it fails to meet the ordinary tests of truth. Those who critique it run the risk of being obliterated. How can a religion that claims that its prophet came to the entire world restrict its miracle to a language that is not spoken by the vast majority of the people of the world? How can a man whose own passions were so untamed gain the right to speak moral platitudes? I've written about this and other critiques of Islam elsewhere. An honest Muslim open to considering these things will readily see that the "God" of the Koran is not the same God spoken of in the Old and New Testaments and that the edifice of Islam is built on a geopolitical worldview masquerading as a religion.[3]

Christians are keenly aware of the threat of secular humanism and its relativist ideology as it rabidly attempts to undercut our religious liberties and continues to draw away our youth. Most Christians see no problem with my defending the faith against evolutionists and new radical atheists such as Richard Dawkins and Sam Harris, but they somehow think that speaking out against Islam is not politically correct. Hear my warning, brothers and sisters, the great battle of our age will not be fought against communism or Nazism; it will be fought against Islam, and I am not alone in this belief. Zacharias has a very important insight into this matter. Hear his prophetic warning:

> Islam is willing to destroy for the sake of its ideology. I want to suggest that the choice we face is really not between religion and secular atheism, as Sam Harris, Richard Dawkins, Christopher Hitchens, and others have positioned it. Secularism simply does not have the sustaining or moral power to stop Islam. Even now, Europe is demonstrating that its secular worldview—one that Harris applauds—cannot stand against the onslaught of Islam and is already in demise. In the end, America's choice will be between Islam and Jesus Christ. History will prove before long the truth of this contention.[4]

Hear my warning, brothers and sisters. If we do not prepare ourselves to meet this coming clash, we will be the victims of our own myopic apathy. The Koran teaches that Allah's goal is to kill every unbeliever in order for Islam to dominate the world. That is the underlying root of Islamic terrorism that the Western nations are blind to. It is a religion like no other religion in the world. Our ignorance will become our demise.

> [B]ut God was desiring to verify the truth by His words, and to cut off the unbelievers to the last remnant.
> —Koran, Surah 8:5

The militaristic nature of Islam is motivated not only by the expectations of the 72 beautiful virgins that await them in paradise should they die during their jihad, but also by the fearful expectation of Gehenna (hell, the Lake of Fire) for those who refuse to fight.

> O believers when you encounter the unbelievers marching to battle turn not your backs to them. Whoso turns his back that day to them, unless withdrawing to fight again or removing to join another host, he is laden with the burden of God's anger, and his refuge is Gehenna—an evil homecoming.
>
> —Koran, Surah 8:15–16

The Muslim is taught to fight the unbeliever until Islam is the only religion in the world. Any religion that opposes this global mandate is considered to be persecuting God and must be obliterated.

> Say to the unbelievers, if they give over He will forgive them what is past; but if they return, the wont of the ancients is already gone!
> **Fight them, till there is no persecution and the religion is God's entirely** (emphasis added).
>
> —Koran, Surah 8:39–40

This does not mean that every Muslim believes in this apocalyptic and violent interpretation of Islam. There are many peace-loving Muslims who do not ascribe to this terrorist ideology. But they seem to be the exception and not the rule. And their power to counter this radical jihadist movement is fraught with great peril as they risk their lives to speak publicly against this barbarous worldview. To these men and women, I have the greatest respect as they risk their lives in defense of women's rights and peaceful coexistence with other religions.

The Nazi crimes committed against humanity were done in the name of the Fuehrer. But this new form of Islamic supernationalism asks its adherents to conquer the world, not for a human being but for Allah. Those who die in the effort are promised a direct ticket to paradise. This religious component is infinitely more dangerous and sinister than the Nazi threat ever was.

But at the other end, the naturalists are promoting the same tyrannical global government through the clever lies of collectivism. Both alternatives lead to tyranny and a monolithic cultural mandate. Both alternatives lead to the oppressive centralization of power. Both alternatives will spell the death of individual freedoms as the foreboding darkness of the coming night inches ever closer on our horizon.

On one side, we have the naturalists promoting the slow track of socialism through economic and political syncretism, which will lead to a socialist New World Order. And on the other side, we have the fast track of Muslims promoting violent supernationalist military conquest in order to bring the world under Sharia law.

Some may argue that the Muslim imperialist agenda does not fall into the category of supernationalism since they promote a religion over the secular governments of the nations. But they fail to realize that Muslims consider all religious adherents as members of the nation of Islam. There is no division between religion and politics in Islam. Their form of government may be a theocracy, but it is nonetheless a supernationalist political agenda. They do not recognize any other national governments as legitimate. As hard as it may be for Western minds to understand this, to radical Muslims, political control over all nations is their religious duty.

As Christians and Jews, we must be cognizant of the fact that God instituted national governments to prevent the centralization of power that invariably leads to tyranny and gives the Enemy of Man the throne over this earth. To absolve the national distinctions in order to create a one world order, no matter how lofty one thinks their goals are, is to give Satan dominion over the planet and to rebel against God's judgment at the Tower of Babel.

Collectivism

Most globalists in our Western culture who promote a New World Order seek to advance collectivism as the answer to the problem of global peace and unity. For this reason, nationalism is being attacked on every front. But collectivism is just as ruinous to the welfare of our planet and to the liberty of its people as the zealous overexaggeration of nationalism, which corrupts it into supernationalism.

In the past, supernationalism has been the overriding historical impetus for colonialism throughout our planet. It championed a form of globalism called imperialism that also invariably ends in tyranny. Following the technological advances in shipmaking after the Enlightenment, European nations spread throughout the planet, dividing up the spoils as they ruthlessly subjugated the less technologically sophisticated people. Imperialism led to enormous injustices and the cruel theft of the natural resources of the nations they carved up in Africa, the whole of America, India, the Middle East, and the Far East. Borders were artificially drawn by the European conquerors to divide the distinct native populations into opposing groups that could then be politically manipulated according to their ulterior motives.

Whether the elite supernationalist group is composed of a particular ethnic group, a religious body, an economic cabal, a single tyrant, or a single country is immaterial. In the end, those who are not members of the elite group are considered slaves to the imperialist elite in power. Colonists were looked down on as second-class citizens whose only purpose was to enrich the conquering nation. This is not true nationalism as it was designed but rather a bastardized and profaned version called supernationalism.

Sadly, some of these nations claimed to be Christian nations, and collectivists and Muslims have effectively used this deception to spur on their own globalist ideology. There was nothing Christian about their imperialist colonialist mentality. This unfettered greed stands as the very antithesis of the Judeo-Christian worldview. The subjugation of another people for cruel exploitation purposes is the very antithesis of God's will for humanity.

But at the same end of the spectrum as the supernationalists are the collectivist promoters of another form of globalism champion, an equally unbalanced ideology that leads to the devaluation of the individual human being (hypoindividualism) and the centralization of power.

Collectivism invariably leads to the same tyranny created by the centralization of power and the loss of the distinction of the individual. It leads to a worldview in which reality is conceived as a unified whole without qualitative distinctions.

Collectivism leads to monism, or an absolute uniformity without distinctions. It is an evil and oppressive worldview. But collectivism and monism are corrupt forms of true truth. Collectivism is a corrupted form of community, while monism is a corrupt and profane version of unity. Monism is an evil doctrine designed to oppress our personhood and rob us of our uniqueness as special creations made in the image of the creator. And collectivism accomplishes the very same thing through political force, violence, and intimidation. It seeks to destroy our individuality for the sake of universal conformity. It is, in fact, the attempt of the elite in power to manipulate the masses, and it is usually disguised and justified as an action necessary for the good of the collective.

But society at large is best served when the constituents within it are free to exercise their individuality. Each individual character then adds a needed dimension that provides balance to the whole and allows the true collective to flourish properly.

What good is a body in which every organ is exactly the same as all the others? Can a body function with only legs or only eyes or only mouths? Each organ adds a needed dimension to the whole and makes the body functional and complete. But is the heart more important than the brain? Can they survive without each other?

While there are distinctions in function, there is no distinction in value. Therefore, each individual in a society should be given the freedom to express his or her individuality, while all should be regarded as equally valuable regardless of any superficial distinctions that can be made in our human family.

The same principle applies to nations. A balanced or healthy nationalism, when functioning properly, should then foment harmony. And in the very same way that individuality enhances the collective, individual nations, which collectively reflect the collective individual character of their individual citizens, it creates a true and balanced nationalism that enhances the global community.

On a personal level, the stifling of the freedom of individuality is a direct affront to our personhood and the way God designed us. In the same way and for the same reasons, the globalist agenda to dissolve nationalism altogether, despite the good intentions of many, is an extrapolation of this myopic and erroneous mentality. At either end of the spectrum, humanity suffers, both individually and collectively, and only an elite group profits.

The supernationalist-globalist erroneously proposes a prideful, egotistical, egalitarian hypernationalism that attempts to impose a singular tradition on all humankind. The collectivist-globalist, on the other hand, seeks to stamp out the diversity in humanity, promoting an equally destructive hypoindividualism that, in the end, also imposes a single tradition on all humanity. Both end up in severe repression and the loss of individual freedoms.

They both attempt to create cultural homogeneous conformity by military, religious, or economic coercion. In any of these views at this end of the spectrum, the common person is reduced to a slave caste, and both individualism and individual rights are severely repressed.

Both of these erroneous counterfeits are like a cancerous cell in the body that takes over all the other properly functioning cells and eventually causes the organism to die. The cancer cell aggressively reproduces itself and feeds off the body, consuming all the other diverse cells that create a living organism. Living organisms survive by the nature of the harmony produced through the individual functions of their diversely created cells.

But when the uniformly apparent cancer cells attack the body, they destroy the other cells. The cancer cells reproduce at the expense of the other individualized cells until the uniform cancer cells have destroyed the proper functioning of the various organs in the organism it has invaded.

This process continues until the body cannot function anymore and simply dies. In the end, the reproduction of the uniformly apparent cancer cells overwhelms and destroys the individual cells, consequently inhibiting the diverse functions of the body that give it life. And so the body simply ceases to function and dies. This is also what happens to collectivist nations.

The danger to the welfare and liberty of all people not only comes from supernationalism but also collectivism. We the people must remain leery of those who espouse concern for our welfare in the name of progressivism, socialism, and collectivism. They are the enemies of our individual liberties.

The collectivist ideology is rather like the hungry master cutting off his dog's tail and, after eating the meat, throwing him the bone. The uncritical mind does not consider the origin of the bone. In this same way, socialism, progressivism, and communism forcefully take from us to return only a fraction. By the time the bureaucracy is finished gleaning our confiscated contribution, all that is left is the bare bone. The uncritical mind pitifully wags the stump at the supposed generosity of its master!

Our nation was wise enough in the last century to see the evil of the supernationalist agenda that rose to power in Germany. And going against our previously entrenched, traditionally provincial, or isolationist mentality, we bravely jumped into the fray during the First and Second World Wars.

But we in the West are now being hoodwinked into accepting one of two equally destructive worldviews: collectivism disguised as progressivism or socialism; or supernationalism disguised as a religious veneer (Islam). In both these extremes, the freedom and dignity of the human being is vanquished for the profit of the elite in power.

Greed is the underlying cause of this calamity that supplants the God-given value of each individual and brings tyranny to the human family. Will we be wise enough to discern these new threats to global peace?

It is the wickedness of greed that inevitably spawns the human tragedy of war. How sad that humans can stoop to such a level and justify the senseless slaughter of other human beings in order to attain some selfish material goal. War is but the end result of selfishness and the inevitable product of rebellion to the way of God.

Yet sometimes, war is the only way to stem the tide of totalitarianism and oppression. We must have the courage to stand firm and unflinching when all other options have been exhausted. We must have the foresight to exercise the wisdom to use force against evil when evil becomes aggressively violent and threatens the lives of those it wishes to vanquish.

Evil must be confronted. It cannot be left unchecked! If it is not opposed, it will expand like cancer and consume the whole. But our opposition must exhaust all peaceful means before resorting to guns and bullets as a defensive measure. We must battle for the minds of all human beings.

Humanity must learn to sift through the clever smokescreens of tyrants. We must not allow our leaders to bring us to the battlefield in search of plunder and economic gain, touting the banner of supernationalism. Our citizenship is first in heaven, and we are to elevate God's principles that bring justice above all else.

Our nationalism must never come into conflict with true justice, for then, it has become profaned. It is our greed that ever brings us to the sound of the gun. It was greed that brought upon us the first global war, which we naïvely thought to be the war to end all wars. My grandfather, Henry Magginnetti, served as a lieutenant in the U.S. Army during World War I. The brutality and carnage he witnessed was unspeakable. As a child, I would ask him to tell me the things he saw. His face grew taut as he shook his head and said, "No, they are too terrible to even remember. It was supposed to be the war to end all wars."

The War to End All Wars?

by
Henry Patiño

Shattered tree stumps in a misty gray hue
Fields of mud, rotting corpses black and blue

Now the callous call to charge comes through
As mindless men do trample morning dew

Running headlong, heedless into mowing walls
Of bullets tearing flesh amidst No Man's Hall

How could man, so blind, so trivial be
To fight for things, which you can see

Falling, screaming into muddy sea
Broken, bleeding, yet they cannot flee

There amidst the corpses do their bloods get mixed
With soil and rock and bayonets fixed

Whizzing death over heads declares
None should lift a head or dare

So many brilliant minds now shattered
Blown to fragments; what senseless splatter

Futures stolen from the bright and young
For the glory and the greed of those beyond

Warm and cozy in their smoked filled rooms
Far from danger and this morbid gloom

When will sense rule man's minds?
When will nations learn to be more kind?

Alas, our hopes of peace seem almost gone
Amidst the roaring and the smell of guns

And yet our hope and faith still clings
Our hearts in hopeful chorus sings

For Shiloh'll come to bring us peace
And sin and war will one day cease.

It was greed that brought about both the First and Second World Wars and every other war that humanity has fought. And it will also be the reason humankind will be brought into the third world war that will usher in the global government of the Antichrist. We must uphold justice no matter the power of the forces arrayed against us. Here, the true courage of a person is tested. Can he or she stand with integrity and resist evil, even at the cost of his or her life?

Many people have chosen this path less trodden in human history and have paid dearly for their bravery. Some lost all they had worked for their entire lives. Others lost their families, and some were tortured, raped, and imprisoned.

And yet when their bodies were fully manacled and to all outward appearances they were utterly defeated, their hearts were freed. For as Solzhenitsyn so insightfully explained, tyrants rarely understand that when all has been taken from a person, their power over that person ceases, for they are free once again. In a fallen world, that is the unfortunate price of freedom.

For now, terror is the norm of life in our rapidly wilting planet. But the time will come when nations will be judged and expunged from the face of our planet. And in that blessed day, humans will not cause terror anymore. Terrorism has become the byword of this age as the time of lawlessness draws near at the end of the second earth. But when the Righteous One comes to end the iron rule of the oppressors, there will be no more terror.

> The LORD is King forever and ever;
> Nations have perished from His land.
> O LORD, You have heard the desire of the humble;
> You will strengthen their heart, You will incline Your ear
> To vindicate the orphan and the oppressed,
> So that man who is of the earth will no longer cause terror.
>   —Ps. 10:16–18

Nations war against nations for only one reason: to obtain financial or economic gains and power. To be sure, these are often masked with religious overtones and utopian illusions. But they are only smokescreens for their underlying greed.

There are few legitimate reasons to go to war. War is morally legitimate in either self-defense or the defense of the oppressed. Unfortunately, that is sometimes a necessity. But we must understand that war is in every case brought about as the culmination of greed from one side or both.

If we abandon the Judeo-Christian ideals and accept a purely relativistic foundation for our morals as a nation, then it is safe to say that greed and avarice will rule the minds and hearts of our people. And the end result of greed will always be war and violence. It is the natural outworking of the evolutionary mantra, the survival of the fittest. Violence is the only underlying matrix of the evolutionary paradigm.

Socialism, Communism, and Justice

The economic failure of communism became evident to most of humanity with the fall of the Soviet Union and the economic metamorphosis of the Chinese government to free market principles. But collectivism is not dead. It is alive and well in a more palatable form called progressivism or socialism. Progressivism, socialism, and communism are collectivist ideologies whose foundational premises are the same.

They all fail miserably to provide a viable substratum from which a just society and a functional economy can develop. Without a doubt, the initial strength of the collectivist movement in every instance has come from the natural reaction toward the heartless abuse of the working class (proletariat) by the wealthy ruling class (bourgeois). It has been the heartless abuse of the poor by the rich that provided the understandable incentive that has fomented many bloody revolutions in much of our world.

The ideal of longing for social justice is, in fact, a proper and just sentiment that is inherent in the *Imago Dei*. It is because each human being has been created in the image of God that humankind possesses an intrinsic desire for justice. It is rooted in the foundation that we have infinite worth and that our fundamental human rights are inviolable because they are God-given.

Communism and socialism borrow from Christianity the ideal of justice, while they fail to provide the philosophical substratum necessary to even legitimize the concept of justice. If they begin from an atheistic presupposition, justice is whatever the state says it is. And although their motive may be often justified, from the viewpoint of the Judeo-Christian worldview, the resulting government provides no justice for the proletariat as a result of the inherent failure of their basic philosophical presupposition to declare any absolute notions of just laws.

If all truths are relative, then the state becomes the arbiter of truth and justice. The individual has no intrinsic rights. The citizen becomes a utilitarian tool for the purpose of supporting the state. In the final analysis, it only results in the transfer of power from one elite group to another. Pragmatism and not justice is the matrix of collectivism in all its forms.

If there is no God, as the atheistic philosophy of communism, progressivism, and socialism stipulates, then there can be no measuring point or basis from which any form of absolute justice could be determined. Under their atheistic presupposition, truth and justice are not absolute. All truth and therefore any resulting notion of justice are absolutely non-binding or relative. Consequently, power becomes god. In other words, the concept of justice under the communist system becomes what is pragmatically profitable for the state since the state becomes god.

For this reason, communist governments, which begin with their naive adherents' idealistic motive to bring justice to the working person, never accomplish that end. And the working class is invariably trapped in a tyrannical state whose prerogatives outweigh the rights of the individuals within them. The concept of the dictatorship of the proletariat is just another smokescreen and illusion used to fire up the masses to do the will of the elite in power. The most powerful argument against communism is communism in practice.

Communists openly deny the existence of the supreme creator and therefore insist that all truth and morals are relativistic. And yet they cannot escape from the notion that there ought to be justice. But if truth is relativistic, then it cannot be universal and absolute. No matter how they try to evade it and try as they may, even their leaders cannot escape the fact that in order for justice to exist, there must be universal truth.

They claim that their battle against capitalistic imperialism is based on their fundamental notion that imperialism is unjust. I wholeheartedly agree that imperialism is unjust. But if there is no

God, then the notion of injustice is simply an illusion, for it cannot be based on any universal truth. They have no grounds to call anything unjust.

Hence they live in a dichotomy. While they claim that universal truths do not exist, they live largely as if they did. And even their leaders sometimes slip and admit that there are universal truths, which they claim, of course, justify their revolutions.

Mao Tse Tung is a prime example. In his famous speech "On the Ten Major Relationships," he attempted to forge an independent communist path that was more in line with Chinese culture than the Soviets and inadvertently slipped:

> We put the problem in this way: the study of universal truth must be combined with Chinese reality. **Our theory is made up of the universal truth of Marxism-Leninism combined with the concrete reality of China.** We must be able to think independently (emphasis added).[5]

Just how Marxism and Leninism become a universal truth in a relativistic system is a bit of a mystery to any thinking person. But most are blissfully ignorant of this oxymoron.

The concrete reality of China is that there is neither religious nor political freedom. Individuals must abandon their God-given human rights to worship according to their consciences and accept the state religion of atheism or pay the brutal consequences. The repressive persecution that relentlessly targets Christian pastors in China overshadows even the infamous persecution of the early church under the hands of the megalomaniacs who ruled the Roman Empire.

Ask the students in Tiananmen Square about the concrete reality of China. According to the Chinese Red Cross, 2,000–3,000 people were brutally killed by the communist government when students, intellectuals, and labor activists waged a peaceful demonstration between April 15 and June 4, 1989, demanding broader freedoms. This is the concrete reality of a communist government that is supposed to be defending the rights of the people. They just mowed them down with machine guns and ran over them like vermin with tanks. How is that construed as the rule of the proletariat in any way?

These underlying presuppositions of the philosophy of communism or socialism are, in fact, corrupted counterfeits of the true communal principles of the Judeo-Christian worldview and the notion that all people are created equal and are therefore endowed by the creator with certain basic and inalienable rights. It is only the Judeo-Christian foundation that can provide a substratum for the promotion of true social justice.

But these so-called universal truths of Marxism-Leninism are simply reflections of the mirage of self-healing that have been propagated by the Great Deceiver. They are appealing to people because of their intrinsic natures as the image bearers of God. People intuitively know that the exploitation of humans and the injustices it produces are wrong.

Marxists therefore capitalize on our human desire to bring justice and equality to all humankind, but their method of attempting to achieve this supposedly just form of government for all people never accomplishes that desire. Moreover, these Marxist principles were founded on misapplications of Judeo-Christian truths, albeit oddly enough the proponents have mostly been atheists and not a few dabbled in the occult, a fact that may offer the inquiring mind another clue as to its real author!

Communism is supposed to be the ultimate in the practice of community. The community practiced by the early Christian church is often used as propaganda to promote the communist ideology.

But the true concept of community is based on both compassion and free will. That is, men and women who have enough to share with those who do not are encouraged to do so. But that is never under compulsion or coercion.

It is their choice to do as they will with the possessions that God has blessed them with. The choice to give is, of course, then rewarded by God. Nevertheless, it is the choice of the individual to give or not to give. And it is as well the individual's choice to decide what amount to give and to whom to give it.

Many collectivists improperly use the New Testament to promote their false ideology of collectivism in their propaganda. But they do so by taking those principles completely out of context.

The early church exercised enormous faith so that many gave all of their possessions to the church for the furtherance of the gospel to all humankind. The critical need to bring the message of redemption through the death of the Messiah to all humanity, from Judea to the farthest corners of our planet, brought many brave souls to the challenge, and their mighty faith will reap the eternal harvest of souls that their gifts made possible. But that was never a requirement or a standard policy that was enforced by the early church in Jerusalem or any other place where it expanded.

To be sure, the hoarding of wealth by the greedy and the unjust exploitation of the working person by those who blindly seek to amass their wealth at the cost of the unfortunates they exploit is inexcusably wrong and unjust. This is most definitely not legitimized by the Judeo-Christian principles taught in scripture, as some communists have tried to claim.

As a matter of fact, Hebrew law required that land would revert to the original owners every 50th year—the Year of Jubilee. That mandate from God was designed to secure the proper distribution of wealth in Israel and to keep the greedy from eventually hoarding all the wealth of the land. Unfortunately, Israel did not comply with God's command, which was one of the reasons the Israelites were taken from the Promised Land during the Babylonian captivity.

The culprit is therefore not the ideal of the right of personal possessions, as the communist would argue. But instead, it is the unjust hoarding of personal possessions and the unjust exploitation of those disenfranchised in order to selfishly amass wealth and power. Greed is and has ever been the ultimate culprit that foments injustice.

The collectivists attack capitalism by claiming that it is fundamentally being fueled by greed and the desire to attain more wealth. They attribute the idea of collectivism as a more altruistic, selfless, and lofty ideal. Contrary to the collectivist claim, the ideal motive that should stimulate capitalism is not greed. It should be a proper desire to provide for your family and beyond to your extended family—your community. But your community is not limited to your immediate geographical location. It extends outward to your nation and beyond to your global family at large.

The collectivist fails to recognize that the unlawful theft of one man's honestly earned goods to give to another is, in fact, the result of the greed of the ones who have not worked to attain those goods. Their self-imposed horse blinders prevent them from seeing the beam in their own eye while they criticize the speck in their neighbor's eye. But in addition to their greed, collectivists must at their basic foundation use force and coercion to steal what does not belong to them in order to redistribute it among those who have not earned it and therefore do not deserve it.

The acquisition of wealth is good when it is sought justly to provide for our needs and to help our neighbors in need. **What government needs to secure is the freedom of all people to have equal opportunity to do so. True justice can come only with true equal opportunity for all citizens in a nation.** This is the only balanced form of economic incentive for capitalism from a Judeo-Christian perspective and by necessity opposes the monopolies of supercapitalism.

Supercapitalism is a corruption of capitalism that bypasses the boundaries of protection that a nation ought to have to ensure equal opportunity for all businesses. It is the greed of the supercapitalist counterfeit system that brings forth the unjust hoarding of wealth and power and the heartless exploitation of others in order to accumulate such inordinate wealth, which government must stop.

Both collectivism and supercapitalism use coercion and force to rob others of the right to earn their wealth. The attainment of wealth is not evil if it is done honestly without unjustly abusing others in the process. It is not wealth that is evil but the greed for hoarding wealth by exploiting others, which is condemned by God.

Hence, the scriptures say that "the love of money is the root of all evil" (1 Tim. 6:10 KJV). It is not money that is the root of all evil, but rather the love of it that brings all the attending problems. Greed is a symptom of an underlying selfishness.

Money can be used to accomplish great good or great evil, and it is neither good nor evil intrinsically. It is our greed for money that causes the unjust exploitation of the less fortunate who are heartlessly abused in order to extract their last possible farthing, as well as their blood and sweat.

The true measure of justice in a nation is the size of its middle class. Those nations where a handful of families own all and the rest are peasants show the end result or product of supercapitalism and their monopolies. The elite bask in wealth while the rest suffer hunger. That does not please God. But make no mistake, there is One in heaven who sees all and judges righteously. The greedy merchants of monopolies will eventually pay for their crimes and indiscretions. The meek will inherit the eternal earth.

But the citizen also has a responsibility to work. When sin entered our world, God said to Adam, "By the sweat of your brow you will eat your food" (Gen. 3:19 NIV). Those who are capable but refuse to work and expect the government to support them through a socialist welfare state are, in fact, in active rebellion toward God's mandate in the Garden of Eden.

Unfortunately, many within the downtrodden masses have un-critically accepted the socialist-progressivist heresy out of desperation in an attempt to climb out of the injustice of their imposed inequality by those who have callously imposed supercapitalism upon them. But socialism, progressivism, and communism do not provide an adequate

THE CENTRALIZATION OF POWER

answer, either. They are each just another blind alley manufactured by
the master deceiver. Both collectivism and supercapitalism end with
an elite in power and the masses as their thralls.

The collectivist heresy is rarely practiced in the ideal form that
is expounded in theory. Once its proponents come to power, there is
simply a change in the elite, and the downtrodden remain abused—a
fact that has been cleverly depicted by George Orwell in his well
known book *Animal Farm*.

But even if its principles were completely put into practice,
it would still be an economic failure that would result in unjustly
keeping those who could better their stature through hard work
from achieving a more prosperous life. At the same time, it unjustly
rewards the unmotivated lazy leaches of society, thus ensuring that
they never become productive members of a community.

Why should I work so hard to earn money if nine months out of
the year I am working to give my money to someone who chooses
not to work and has become a ward of the state? What motivation is
there to work harder when there are no financial rewards to pass on
to their children?

That is not to say that we should not have compassion for those
who are less fortunate. But that compassion should be measured
against the willingness of that individual to produce at whatever level
he or she is capable. The scriptures say that "if anyone is not willing
to work, then he is not to eat, either" (2 Thess. 3:10).

So from a Judeo-Christian perspective, all of us should be
given equality of opportunity but not absolute equality. Rewarding
laziness creates a negative pull on the overall economy of a nation as
well as an unjust theft of the rightfully earned rewards of another's
labor. The enormous growth of the welfare state has not helped our
economy one bit. Neither has it helped those disenfranchised from
climbing out of their poverty. It is an artificial stopgap that produces
no long-term benefits. It keeps the disenfranchised in slavery to the
government dole. It solves nothing.

We must help those who are truly in need. But those who can work must work. In this sense, then, it is more appropriate to divest our national energies in workfare rather than welfare. The government has a vested interest in providing jobs for its people but not in subsidizing laziness. Giving them a job where they can be a positive influence on the economy and not lose their self-respect should help the people who are out of work.

But the best way to give them work is to provide the freedom and conditions for free enterprise to thrive. This is the engine that fuels prosperity and provides jobs. When the government inhibits small businesses from competing by overburdening them with taxes and unjust regulations, they cannot compete. Workers lose their jobs, and the economy tanks.

The progressivist economic policy of burgeoning federal governments destroys the working class by destroying the small businesses that are incapable of thriving under such massive taxes. They may in the short run provide a few more government jobs, but it is unsustainable. The federal budget becomes so gargantuan that taxes can no longer uphold it. It is doomed to failure. It is doomed to bankruptcy.

The taxes levied on the working population to bankroll these parasites of society rob people of any hope to better their lives and their children's lives by taking from them the capital they need to invest and prosper in the hope of improving life for their progeny. I dare say that the feudal lords of the Medieval Age allowed their serfs to keep more of their earnings than our modern socialist governments allow their citizens to keep today. Think about that!

For the most part, I suspect that these socialist systems are designed by the global financiers to drive the nations into greater national deficit. That, in turn, ensures greater economic control of these governments by those international financiers who are able to bankroll their enormous deficits.

The naïve notion bandied about by idealists that socialist governments take care of their people against the desires of the rich is preposterous beyond measure. It is the very rich moneylenders who profit most from socialism and progressivism. We can cite the national debt of Greece as a prime example of this undeniable reality.

For years, the Greek government has spent more than it could receive in taxes. The United States is following in Greece's footsteps. Our national debt is increasing at an average of $2.5 billion per day. You do not have to be an economist to see the handwriting on the wall. When that bubble bursts, our nation will be in economic ruin, and the future will be dark, indeed.

The problem intrinsic to the socialist ideology of giving people the promise of free stuff is that the stuff is never really free. The Greek national debt continued to increase without any attempt by the government to rein in its spending. It is the same here in America where Congress continues to raise the nation's debt ceiling indiscriminately. Folks, kicking the can down the road is a fatal policy that will sooner or later explode in our faces.

In 2015, the national debt in Greece reached a whopping $356 billion that was owed to a variety of banks, including the International Monetary Fund (IMF). That deficit was 177% higher than its gross domestic product (GDP), making it absolutely impossible for Greece to pay even the interest on the debt. Adding fuel to the fire, the GDP in 2015 had actually fallen 25% since 2010. Middle class small businesses had gone bankrupt. As a result, Greece's unemployment rate was 26% of the population.

But Greece's problem did not begin in 2015. When the manufactured global market crashed in 2008, the cost of borrowing money conveniently rose massively. In 2010, Greece had received a bailout sum from the IMF of 240 billion euros, but on June 30, 2015, Greece failed to make the 1.5 billion euro payment to the IMF. And that is when all hell broke loose, and the country was close to total bankruptcy.

The IMF, without firing a shot or sending a soldier to conquer Greece, basically owns the country outright. Greek socialist policies have now effectively condemned the future generations of Greeks to a stark future of economic slavery to the IMF. We are on that same road, and so are most European nations that have adopted the same suicidal socialist ideology. Portugal and Italy are but a few steps behind Greece.

There are several major policies held in common by socialist countries:

1. The government begins to take over private industry until the central government controls all industries.
2. Taxes increase tremendously in order to pay for the expanded and bloated federal bureaucracy and their social welfare programs.
3. The middle class is basically taxed out of existence, and the country begins to spiral upward in its national debt exponentially.
4. The quality of healthcare diminishes drastically, and the industries lose their incentives that are the natural by-product of competition in the free market systems. Thus, innovation suffers greatly.

And what about Sweden? Many socialist Democrats love to point to Sweden as a model of how socialism can work. Let me share a January 2, 2019, article by John Stossel in *Capitalism Magazine* titled "Sweden Isn't Socialist." It provides the real story behind Sweden's capitalist foundation.

For years, I've heard American leftists say Sweden is proof that socialism works, that it doesn't have to turn out as badly as the Soviet Union or Cuba or Venezuela did.

But that's not what Swedish historian Johan Norberg says in a new documentary and Stossel TV video.

"Sweden is not socialist — because the government doesn't own the means of production. To see that, you have to go to Venezuela or Cuba or North Korea," says Norberg.

"We did have a period in the 1970s and 1980s when we had something that resembled socialism: a big government that taxed and spent heavily. And that's the period in Swedish history when our economy was going south."

Per capita GDP fell. Sweden's growth fell behind other countries. Inflation increased.

Even socialistic Swedes complained about the high taxes.

Astrid Lindgren, author of the popular Pippi Long-stocking children's books, discovered that she was losing money by being popular. She had to **pay a tax of 102 percent on any new book** she sold.

"She wrote this angry essay about a witch who was mean and vicious — but not as vicious as the Swedish tax authorities," says Norberg.

Yet even those high taxes did not bring in enough money to fund Sweden's big welfare state.

"People couldn't get the pension that they thought they depended on for the future," recounts Norberg. "At that point the Swedish population just said, enough, we can't do this." (emphasis added).[6]

Sweden did an about turn, reduced its government bureaucracy, and abolished government monopolies. They cut their public spending drastically and privatized the national railroad. They abolished the inheritance taxes and sold their state-owned businesses, such as Absolut vodka. And, they reduced their unsustainable pensions. The result is that Sweden made a complete economic turnabout and became one of the world's richest nations.

The Swedes astutely realized that the free market principles that bring competition were essential to make their economy prosper. Socialism did not work in Sweden, and they wisely took steps to correct it. They privatized because they realized that these government monopolies without competition failed to provide for them the innovations that make industry prosper. Sweden switched to a school voucher system that forced public schools to compete with private schools and improve the quality of their education, and they now have private pensions that supplement their equivalent of our Social Security system. Rather than being an example of a successful socialist system, it is an example of the failure of a socialist system and the efficacy of the free market capitalist system.

Inevitably, governments faced with insurmountable debt due to their enormous social welfare programs turn to the printing of money, which causes runaway inflation. As a consequence, it reduces the ability of the middle class to that of the disenfranchised. Not only do their life's savings for retirement and for the betterment of their children's future disappear in a puff of smoke, but by devaluing the currency, the middle classes in these nations are no longer financially capable of mounting any concerted resistance to the authority of the oligarchs or autocratic despots in power who are shoving socialism down their throats. That gives the very rich complete tyrannical control of these nations.

If America is to remain a free nation with a thriving middle class, we must reduce the size of the federal government and force our politicians to balance our federal budget before the dollar implodes. That must be a high priority for each and every American family.

Absolute Equality Is Unjust

Progressivism, socialism, and communism are collectivist ideologies that adhere to the principle of forced, absolute social and financial equality. The fundamental collectivist ideology of an absolutely classless society is, in fact, also cruel and unjust, for it destroys the distinctiveness of the individual. The diabolically clever ploy used to deceive the masses is the promise of a utopian classless society with unconditional social and economic equality.

Certainly, anyone with an ounce of moral conscience abhors the wickedness of forced inequality. We have seen the ugliness of slavery and prejudice as it scarred our human race. Forced inequality through the tyranny of the elite over the masses has ever plagued our human family. At first glance, the idea of unconditional social and economic equality sounds like the proper thing to do. After all, who can be against equality? But absolute equality is as unjust as forced inequality.

Only the Judeo-Christian position that fundamentally stipulates that the Almighty created all people as equals can provide the proper substratum from which the concepts of true equality and justice can be enforced. If there is no God, as championed by the communists, then equality is a baseless ideal where the survival of the fittest and strongest is the very matrix of reality.

Hence, communism has no real philosophical basis from which to enforce equality. **What is more, communists' idea of absolute equality without distinctions is erroneous and unjust. It categorically ignores justice.**

God is not for absolute equality! That might surprise you, but it is true. God is for absolute justice, which takes into consideration several points. There should be equality of worth granted to all human beings, regardless of race, sex, or age (including unborn humans), and this must work outwardly to equality of opportunity for all. In this, God is absolutely for equality. All human beings have absolute equality in value, but they do not have absolute equality in actions. Our rewards for actions should be based on the merit of the actions we undertake. This is the measure of justice, and it is the central point of attack engineered by Satan.

Humankind has the option of making choices granted to them as created beings. Of all the creatures God created, we alone were created in His image and thus endowed with the ability to make moral choices. That is what makes us beings, reflected in that transcendent characteristic we know as personhood. It is what creates in us the unique ability as individual beings to make moral choices and be a unique individual in the midst of the rest of our human family.

Consequently, this universal and intrinsic ability to choose, which every human has, by natural necessity brings on the individual a consequence that is directly tied to his or her choice. These choices define our moral responsibility and rectitude before God and all humanity. It is this consequence in direct relation to the choices we make that Satan has worked so hard to eliminate in our minds.

That is, he has sought to remove the responsibility for the choices we make. He has attempted to do away with both the positive and the negative repercussions of choices. The artificial uniformity imposed by this kind of absolute "equality" destroys people's free will and attempts to ignore the real and tangible consequences of our choices, not only in the temporal realm but also in the world to come.

We see this reflected in the collectivist ideologies by removing the just economic reward for hard, honest labor and by unjustly rewarding the idle or lazy in a welfare state. Hence, these political-economic philosophies have trampled true justice. Our choices should be rewarded when they are constructive. Furthermore, they should be penalized when they are destructive. This is true justice. This is the true underpinning of the Judeo-Christian philosophical position.

There is, therefore, in God's mind no absolute equality for every human being. For He judges humankind according to the choices they make. And that is absolutely just and proper. God balances righteousness, justice, and equality in order to properly reward our choices. And our governments should also reflect that same justice as earthly representatives of the divine mandate for true justice. For this very purpose they were created.

Justice, then, demands that equality be not absolute. It is justice that must be absolute. If I were treated with the same reward that a monster such as Hitler would be given, then I would say that there is no justice. And if God would reward Hitler in the same manner as Mother Teresa, then we could plainly say that God would not be just.

But such is not the case. God is a loving God, and therefore, He must be absolutely just in His actions. That same just character should be reflected by us personally and nationally as a society.

If in the economic arena I was a hard-working person and, irrespective of the quantity and quality of my labor, was granted the same reward as someone who did no work, I would be unjustly treated. And that is the basic reason the communist cooperatives simply do not work. They are a dismal failure because they fly in the face of who we are. We saw this in the Soviet Union and China as millions died of famine due to this erroneous collectivist ideology.

In all collectivist ideologies, justice is trampled to the ground and the common person is exploited and abused. Therefore, Satan's brand of collectivism is unfair, unjust, and immoral.

Rewards must be based on the efforts that individuals have made since they are given the same opportunity and equality in consideration. It is not absolute equality but rather absolute equality of opportunity.

This is the basic framework of a just, capitalist economic system within the framework of the Judeo-Christian worldview. It is only the application of the true principle of Christian community that provides a balanced collective and does not destroy the distinctiveness of the individual. That ensures justice and economic opportunity for the individual. It provides a just system of financial reward for honest labor that does not depend on the enslavement of a sector of society in order to function. On the contrary, it promotes a healthy drive to be productive and at the same time selfless.

However, if the selfless-community component is removed from capitalism and certain regulations that ensure justice are not implemented, then it mutates to a cancerous supercapitalism. These regulations must be only the minimal amount necessary to make sure monopolies do not exist and equal opportunity is protected. A government that does not secure the rights of individuals and allows the elite to subjugate the masses is as evil as any other form of tyranny.

In the same way, the forced procurement of someone's possessions to be given to others by a third party (as championed by communists, socialists, and progressives) is nothing less than outright theft in the eyes of God. It discourages the individual who is robbed from producing more and encourages the one receiving to not produce at all.

That is the fundamental reason that economies of collectivist governments are a dismal failure. They interfere with the very mechanism that allows for the distinctiveness of an individual to flourish. In fact, they go against the very nature of human beings. They go against the very nature that God gave us as distinct and unique individuals.

Curiously, even communist leaders admit that the majority of the people they lead do not agree with their socialist programs. The very proletariat whose rights they claim to champion do not agree with their state-imposed direction.

So all that remains in these socialist or communist governments is forced cooperation by intimidation and violence. In his Talks at Chengtu in 1958, Chairman Mao gave a speech titled "The Pattern of Development" in which he frankly admitted as much.

> The line of building socialism is still being created, but we already have the basic ideas. Of the 600 million people of the whole country, and the 12 million Party members, only a minority — only a few millions, I fear — feels that this line is correct.[7]

Only a small minority of Chinese agreed with Mao. The progressivist egalitarian mentality condescendingly treats the rest of society as ignorant children who do not know what is best for them and must be forced to comply or die. They deceive no one but their own minds while they entrench themselves in comfort and luxury and enslave the rest of society. Fools are they who myopically settle for the immediate gratification of the senses of promised free stuff and fail to see the future enslavement it will cost them.

The intrinsic problem of the centralization of power in the hands of bureaucratic experts is that it stymies individualism and, in fact, never leads to progress; instead, it leads to repression. No better example can be named than the agronomical practices of the Soviet Union that forced collectivization of Russian farmlands.

When the Bolsheviks first took over, they not only rounded up every person who had ever been involved with the Tsarist government, but they also rounded up all professionals (doctors, teachers, engineers, and anyone with any kind of education) who opposed the collective ideology. Most were executed on the spot. Those who survived were sent to the gulags (concentration camps) to serve as slaves in the minefields of Siberia or dig canals and other infrastructure projects for the Soviet Union.

All owners of large properties were stripped of their land and either killed immediately or interned in concentration camps. Once their power was cemented, the Bolsheviks went after small landowners whom they demonized by calling them kulaks.

> Twenty-five million rustic households would be forced into collective or state farms and those who resisted would either be dealt with by troops or the OGPU—the Unified State Political Directorate—that had replaced the Cheka. In January 1928, Stalin went to Siberia to oversee the confiscation of grain from the independent farmers who were accused of hoarding it. These smallholders, called *kulaks*, were now considered class enemies.
>
> The Bolsheviks called a village farmer who owned more than 24 acres of land or employed farm workers a *kulak*— the Russian for 'fist.' Stalin saw them as the potential leaders of a future insurrection and in 1929 he proclaimed the policy of "liquidating the *kulaks* as a class."[8]

Cheka and the OGPU were the forerunners of the Soviet secret police, later known as the KGB. The collectivist farms were an economic disaster that brought about a famine of unprecedented scope. Adding to the fiasco of this bureaucratic system was the choosing of an agronomist, Tofim Denisovich Lysenko, who claimed to have invented an agricultural technique that he claimed would quadruple harvests by exposing wheat seeds to high humidity and low temperatures. Because Lysenko was an agronomist who wholeheartedly approved of the Soviets' collectivization of farms, he was given the state's blessing.

> In fact, the technique, known as vernalization, was not new—it had been known since 1854—and it did not produce the yields Lysenko promised. However, Lysenko was one of the few agronomists who supported collectivization and while other biologists were conducting heredity experiments using fruit flies he at least was studying cereal production. He was hailed as a genius, a scientist who came up with solutions to practical problems. . . . Other biologists were, by comparison, considered 'wreckers,' because they refused to apply their science to the Soviet economy and their methods, particularly the study of genetics, were seen as inherently 'bourgeois,' if not 'fascist.'
>
> On the other hand, Lysenko rejected orthodox genetics by maintaining that evolution was instead based on the inheritability of acquired characteristics. This concept was central to the framework of agricultural theories that became known as 'Lysenkoism.' Scientists who opposed Lysenko's ideas were arrested, imprisoned, executed or sent to labour camps. In 1948 genetics was officially declared a 'bourgeois pseudoscience' and research was discontinued. Lysenkoism continued under Khruschev, who fancied himself an agricultural expert.[9]

The myopic ideology of atheistic communism that accepted the false evolutionary claim that acquired characteristics could be inherited forced upon the Soviet Empire the false Lysenkoism agrarian philosophy that was, in fact, a pseudoscience. The peasants paid with their lives by the millions through the famines that collectivism created in the Soviet Union. That is the great danger of autocratic centralization of power.

Most of us have seen in biology books the illustrations with the giraffes straining to reach the only leaves left on the treetops. Each successive picture showed an elongation of the neck as the result of a natural evolutionary process. Modern genetic science has shown conclusively that this is an utter genetic impossibility.

We cannot pass on to our offspring any characteristics acquired during our lifetime. Not a single trait can be passed to our offspring unless they are already encoded in the genes from the moment of conception. At the precise moment of conception, the combination of parental genes presets the genetic composition in the offspring of all creatures. Only what the organism received at conception can be passed on since the combination of the two parental genes results in their uniquely encoded offspring.

Yet the myopia of the ruling elite in the centralized regime of the Soviet Union would not tolerate any scientific endeavor that was not officially sanctioned by their collectivist ideology. Such is the rule and not the exception when power is centralized.

But this is not a unique experience in the communist regime of the Soviet Union. Mao caused the exact same mayhem in China when he forced collectivism on his nation. The famines created by the failed collectivist ideology during his cultural revolution claimed millions of Chinese victims. The same thing can be said for Ukraine when Stalin forced collectivism on the people. The toll in the loss of millions of human lives has been etched in blood and suffering into the annals of history for any who want to know the truth.

But the blood of the millions who were killed by famines or purges cannot alone begin to measure the horrors created by the collectivist centralization of power. The measure of the true horror experienced must include the misery created by the oppressive dictatorial regimes that repressed many millions more, whose individual God-given freedoms were completely denied. Those freedoms we hold so dear, such as the freedom of speech, the freedom of religion, the freedom of the press, and the freedom of individual artistic expression, were forcefully repressed, and any who dared to resist were summarily executed or taken to forced labor camps for the rest of their lives. Most prisoners did not survive the cruel conditions in these hellish concentration camps.

Few in the West can begin to imagine what it is like to live continually under the total fear of the secret police. Anyone who showed any individualistic tendencies was considered suspect and an enemy of the state:

> Other people were imprisoned for simply showing 'individualistic tendencies' — a crime in a collectivist society — while artists and writers were sent to the Gulag if their work did not conform to the doctrine of 'Socialist Realism,' as decreed by Stalin. People who had lived abroad or who had relatives living outside the Soviet Union were also liable for imprisonment. . . . The paintings of the French impressionists were removed from galleries and museums and the work of Jewish artists — considered 'rootless cosmopolitans' — were also repressed. Even approved 'Social Realist' artists were condemned if foreign influences were detected in their work.[10]

It is my sincerest hope that the young people in our nation, those who have been labeled Millennials, would thoroughly research the truth of the things I have been warning about. Their future hangs in the balance. The promises of progressivism are all lies. Progressivism will not lead to progress and equality; it will lead to repression, and the horror brought upon that generation will be darker than they can imagine. The choices they make today could lead our nation into a time of darkness that none in our nation has ever experienced before. I beg Millennials to heed my warning before the American dream turns into the American nightmare.

Capitalism and Supercapitalism

The Holy Scriptures do not specifically use the word *capitalism*. It is a modern word that did not exist at the time of the prophets. But its basic premises are in harmony with the scriptures, which declare that working people are worthy of their wages and that they have the right of personal property and venture. Each of the children of Israel was given a portion of the Promised Land as an inheritance.

The idea of hiring workers for the field and treating them fairly is repeated over and over in the scriptures, both in the Hebrew Tanakh and the New Testament. But it does so in a humane and just environment where the employer should not exploit the employee. Moreover, it provides for the very poor by allowing them to take from the field all they can carry out on their own (see the book of Ruth). Justice requires that a just balance exist between an employer and an employee.

We can point to the unscrupulous employers or the robber barons in the past who chained children to the machines in their factories. It is one of many examples of abuse by employers. Many books have been written on this cruel practice, and I need not tarry here. The hired thugs who killed many workers trying to stand up for their rights as employees have marred the history of the industrialization of America. It is a matter of historical record, and I will not elaborate on it other than to say that the establishment of unions was a necessary step in order to bring some justice to working people. But I am of the opinion that our modern unions are now pushing the progressivist agenda that will enslave the masses. They are no longer knights in shining armor; they have become a branch of the oligarchies, puppets that sing the siren song of socialism.

Government's place is to protect the rights of individuals. Corporations are institutions that are also legally defined as individuals. Thus, the government is also responsible to ensure that bigger corporations are curtailed from trampling on the rights of smaller corporations through unscrupulous methods. Likewise, the government should be prevented from accepting favors from lobbyists. This form of crony capitalism is a corrupt bastardization of true capitalism that leads to supercapitalism and away from the intents of our Founding Fathers and the God-given right of equal opportunity for all citizens.

Regulations must exist that prevent large corporations from abusing smaller corporations and becoming monopolies with a few elite who control a nation's entire commerce. That is supercapitalism, and it is an aberration of true capitalism, which is as evil as collectivism. It is not only unfair to smaller corporations but also leads to the eventual extortion of the people of a nation.

When corruption in a nation becomes such that a politician can be bribed to pick winners and losers in the free field of economic competition, then the criminal act is no less evil than choosing which citizen can have liberty and which one cannot. It is an abominable travesty of our intrinsic God-given liberty to compete in the free market.

It is my opinion that the only way to battle such corruption is to create a punishment severe enough to curtail the temptation. It should be made law that all board members of any corporation who are found guilty of bribing a politician should be stripped of their citizenship and kicked out of our borders, along with the politician who took the bribe. That, however, would require politicians to be willing to enact a law that would end their gravy train. Let us just say it would be a miracle of the first order.

On the other hand, overregulation by the government could inhibit the rights of corporations, which could inhibit their growth and directly impact their financial progress that would otherwise result in a growing economy that produces more jobs. Finding the balance between regulations that are necessary to keep corporations honest and those that are unnecessary and overburden their systems with repressive requirements that destroy their ability to make a profit is the task before us.

Overregulations can also destroy smaller businesses that cannot afford a team of fancy lawyers, accountants, and lobbyists to navigate through the red tape. Overregulation unfairly favors large conglomerates and thus artificially winnows out any competition. It paves the path to the centralization of commerce by an elite group of merchants.

When the economy slows down and our nation's gross domestic product suffers, so do its citizens. Not only do they lose their jobs and their homes, but to add insult to injury, they are inevitably taxed more in order to maintain the federal bureaucracy. It is an unconscionable travesty for the working men and women of the middle class and the poor. This is unfortunately the tactic used by the progressivist agenda to impede the practice of free market capitalism by artificially choosing winners and losers.

We witnessed this outrageous practice firsthand with the bailouts of the auto industry and Wall Street firms. I need not elaborate much on this topic. Most Americans were outraged when our hard-earned tax money was given to these criminally negligent corporations that turned around and gave themselves huge raises. The golden parachutes were numerous, all at the expense of the taxpayer— the working men and women who were hoodwinked by these rapacious vultures.

It also creates a burgeoning federal government that is top-heavy with countless government employees who have to enforce needless regulations. Each year, the federal government hungrily devours more and more power from local and state governments in order to centralize all control of banking, industry, telecommunications, transportation, energy, and especially public education (more aptly called public indoctrination).

For capitalism to be successful, there must be as little interference as possible by the government. That fosters confidence to risk capital in the future. An oppressive over-taxation of businesses reduces the capital for new investments and short-circuits the development of new industries and subsequently reduces new jobs.

Regulations must be limited to laws that impede companies from infringing on the rights of competing companies in order to gain an unfair advantage. The rights of a corporation are no different from those of a person. Our personal rights end where the nose of the next individual begins. No single person or select group of persons can be given extra rights in order to unfairly compete with others.

The progressive policy of burdening corporations with extremely costly superficial regulations serve only to (1) destroy the middle class by bankrupting small businesses and (2) push larger companies out of the United States and into underdeveloped nations with cheaper labor and fewer regulations. Only those mega-corporations unfairly chosen above others equally competent by the federal government can survive in such an arena.

Progressives' globalist agenda has brought numerous international trade agreements such as NAFTA, which have negatively impacted laborers in the United States. These international trade agreements have stripped our nation of manufacturing and effectively parceled it out to the rest of the world.

At one time, the United States was the largest exporter of manufactured goods in the world. Progressives have killed our industry with their policies. Now, China and India are reaping the spoils of the manufacturing industry that we once enjoyed.

Moreover, they have strapped small companies with regulatory burdens that keep them from growing. Larger companies can simply move out of the country and begin anew. But smaller companies do not have the funds to do that.

Seventy percent of the jobs in our nation come from small middle-class businesses that are now being choked to death by spiraling taxes and overregulation by the progressives in Washington, DC. Is it any wonder that we hit record levels of unemployment in our nation, which was staggering from the internal blows it has received from the venom of progressivism. Fortunately for us, President Trump's administration has embarked on a deregulation crusade that has begun to reverse this process in our nation. The results are tangible not only in the increase of our GDP but in record-low unemployment for all sectors of our society. It has also resulted in the return of many manufacturing jobs that had fled overseas due to high taxes and overregulations of past administrations.

Congress continues to raise the debt limit, and our national debt is rising out of control. It is practically useless to name the dollar amount since it is rising so rapidly that the number becomes obsolete in days. If we continue at this rate, we will soon no longer be able to pay even the interest on the loan, much less the principle. When that day comes, our dollar will default, and our economy will fail completely.

The savings of retirees will vanish in a puff of smoke as the value of the dollar plunges through hyperinflation. Millions will lose their jobs and then their homes and cars. A wheelbarrow of dollars will not buy a loaf of bread, just as we saw in Germany after World War I.

In that day, I predict the forces that engineered this bankruptcy will come to the rescue with a proposal to establish a global monetary system. They will claim that is the only solution to the problem. But that will be a lie. If we are not able to teach Millennials these fundamental truths, collectivism will rule our world, and history will look on our generation as the one that killed the American dream.

The solution to the problem begins today. Don't spend more than you have (which is not rocket science)! Reduce the obsessive tax burden from the shoulders of those who are working. Free them to grow and employ others. Return our dollar to the gold or silver standard in order to create confidence in the value of the dollar and promote investments. Ensure equal opportunity to all small businesses by providing a single tax across the board to all businesses. Annul all tax loopholes to the large corporations such as General Electric, which paid no taxes last year. Is it a coincidence that the chairman of the board was chummy with Obama?

Capitalism can be corrupted into a malformed and malevolent version, which can be called supercapitalism, an aberrant form of capitalism that has grown to monopolize certain industries. These typically have deliberately sabotaged other competitors in order to establish dominance. Many of the great industrialists and financiers acquired their fortunes in this unscrupulous manner. Again, this is historical record that has been amply documented, and I need not elaborate on it except to say that you can be sure that their recompense will be meted out by God in due time.

It is my opinion that a balanced form of capitalism is in accord with the Judeo-Christian worldview, while supercapitalism is in complete opposition to its basic principles. Supercapitalism is evil, for it callously disregards the welfare of its laborers and consumers and the rights of honest and legitimate competitors.

Supercapitalism is an unrestrained form of capitalism that is malformed and fed by greed. It exhibits itself in a selfish disregard for the right of another to compete in the free market. Supercapitalism is an attempt to ruthlessly stamp out all opposition and establish a complete and absolute monopoly of a given market.

Once a monopoly has been attained, supercapitalism eventually results in an unjust gouging of consumers in that given market. Invariably, it allows the monopoly to artificially hike the prices of its commodity to illicitly increase profit margins without fear of any competition.

The inevitable result is that under supercapitalism, the working class is hemmed in and exploited by the rich. It is, again, the end product of the centralization of power in the consumer market. And invariably in this system, the common person loses his or her God-given human rights and is eventually economically reduced to a slave caste. Supercapitalism thus exploits the laborer and the consumer, while eradicating all or most of the possible competitors.

More often than not and with few exceptions, our human history has shown that the rich have coldheartedly abused the poor in each nation in every generation. But their greed has not ended at the boundaries of their nations. History has documented over and over again those who have heartlessly sought to profit from financing the expenditures of armed conflicts for the booty and the pillaging of other nations.

The blood of innocents caught in the middle of these political-economic machinations by the powerful and greedy has soaked the soil of our planet from the beginning of recorded history. But justice will one day prevail, and not a single act will escape the notice and judgment of the Almighty when the Day of Judgment at long last arrives.

Be that as it may, the fact that some people are wealthy does not necessarily mean they have amassed their wealth illicitly. There are many who use their wealth to do great good for humanity. Those who through hard work and honesty have bettered their lot should not be punished for doing so. Stirring up class warfare into a revolutionary fervor has ever left in its wake a sea of blood and no real solution to the problem. A just balance is God's desire.

Many unscrupulous politicians thrive in maintaining divisions among the citizens in order to promote their own political agendas. Playing identity politics is the tactic of revolutions everywhere. Political agitators and so-called community organizers with a collectivist agenda, including those trained by Saul Alinsky, never really serve the needs of working men and women. The public must not tolerate their seditious activity against our constitutional republic. We must be more circumspect in following leaders who ultimately want to destroy our form of government for a collectivist agenda that will strip all people of their liberty.

Nevertheless, the attempt to unfairly keep the poor from competing in the free market through cartels, monopolies, bribes, and even overt violence should be opposed by a just form of government. When large corporations gain control of politicians, working people are bound to eventually rebel against unfair exploitation.

The great and illicitly created disparity between those who have and those who have not has thus understandably ignited the flame of revolution in the hearts of the downtrodden on our planet. Sadly, this process will be with us until the end of this epoch when the Righteous One returns to bring justice and freedom to all the oppressed.

It is this great injustice that has been effectively and ruthlessly exploited by communists. The ideal of a classless society has become the highly volatile universal fuel for communist revolutions. But communists have never accomplished what they idealistically hoped for. It is an illusion that disregards the fallen nature of humankind.

They erroneously believe that the goal of establishing a classless society without distinctions is just. But that is simply not true. The ideology of a classless society, as we have already discussed, is a corrupted interpretation of the Christian ideal of the equal and infinite worth of every human being created in the image of the creator. But there is a difference between equality of worth and opportunity and absolute social-financial equality. The progressive ideology robs us of the fruit of our labor and is therefore unjust.

Supercapitalism and Socialism

From the Judeo-Christian perspective, the supercapitalist offers no moral advantage. Both supercapitalism and communism are in direct opposition to the Judeo-Christian worldview.

The supercapitalist unjustly hoards the profits of the labor of the masses without regard for their welfare and for fairness in the distribution of wealth created by their labor. The collectivist does the same, amassing wealth for the centralized government and its ruling elite. In the end, both economic ideologies rob laborers of the benefits of their labor.

Supercapitalists typically establish vast monopolies and unjustly keep any competitors from entering the marketplace. That is absolutely at odds with the Judeo-Christian worldview that champions equal opportunity for all individuals, including corporations, and a just compensation for their labor.

Supercapitalism, like supernationalism, is an elitist worldview. It is a corruption of a balanced and healthy form of capitalism. But in either supercapitalism or collectivism, the people are robbed of the deserved fruits of their labor to unjustly overcompensate the elite in power. Both supercapitalism and collectivism provide an economic system that centralizes the profits for an elite few. In the end, there is no difference between the two.

Although socialism in theory stipulates that the people own the nation's industries, in reality it is the government that owns everything. In some forms of moderate socialism, the right to personal property is permitted, but the fruit of the labor is still robbed through inordinately high taxes that put the elite in control of the government, and those who do no labor become parasites of the state.

These are not at opposite ends of the spectrum but rather at the very same end. They are unjust counterfeit shades of capitalism. Capitalism, on the other hand, resides at the opposite end of the spectrum of these corrupted views.

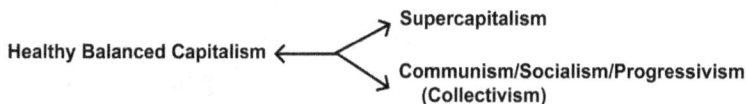

Healthy Balanced Capitalism ← Supercapitalism / Communism/Socialism/Progressivism (Collectivism)

Neither supercapitalism nor collectivism can provide a just and equitable system for the working person that can ensure social justice for the individual and the opportunity for economic prosperity. Both supercapitalism and collectivism create a powerful elite that thrives from the exploitation of the common person, eventually and inevitably destroying the economy. Under these counterfeit systems, the common person is, in essence, invariably reduced to a slave caste.

The economic success of a supercapitalist system may for a limited time function more efficiently than collectivism. There may be some prosperity initially, before the monopolies are cemented, but once they have become entrenched, prosperity will disappear for the common person.

Peace and domestic tranquility are short-lived because they are borne on the back of an imposed economic slavery that will eventually lead to outright rebellion by those unjustly enslaved. In either case, the common person is exploited, and both economic structures are destined to eventual failure.

Some, having become aware of the dismal failure of the communist economic system, have sought to short-circuit the more negative aspects of communism by reworking its doctrines into a more palatable form that includes some capitalistic elements. The term *progressivism* is currently being used to include these more moderate forms of collectivism. But socialism-progressivism is, in fact, fraught with the very same failings as communism. Moreover, its more moderate external veneer does not change the diabolical nature of its internal core doctrines held in common with communism. In fact, a case can be made that they are but stepping-stones toward communism.

The trend toward socialism-progressivism in the world deeply concerns me, as it impinges on the future liberty of all humankind. There is a powerful trend in our global thinking, no doubt supernaturally designed and executed, that is aligning the nations of the world with this counterfeit socialist system invented by Satan. But it seems to me that the European nations, which have been spurred toward this socialist agenda, are simply falling prey to the design of supercapitalists who have much to gain from this.

Supercapitalism can effectively use socialism to gain further control of the world's economies. By their very nature, socialist economic systems create soaring national debt. The more a nation is indebted to the international financiers who fund their federal reserve banks, the more power the financiers are given over the affairs of that nation. And consequently, less power resides in the people to determine the affairs of state.

As the wealth of the middle class is depleted through the soaring tax burdens of a socialist system, their ability to restrain the interests of the wealthy elite diminishes. Socialism is simply a way to bring upon us the iron fist of supercapitalists in due time.

Those young idealists who are brainwashed by our modern indoctrinational system (public education) and who myopically see only the end of their noses will be rudely awakened when the time comes for supercapitalists to make their move. The middle class will disappear and with it the only powers to resist the global tyranny of the supercapitalists that will, in the end, overtake us.

Satan has been hard at work in our society to short-circuit a balanced form of capitalism and nationalism through the greed of the wealthy and the powerful. A corrupted form of capitalism has now risen in our world. That is especially so in the last two centuries as this political and economic system has slowly infiltrated behind the scenes in most Western nations.

The reader must be aware that this perverted version will destroy the true intent of the capitalist, free enterprise system. I am speaking of the smokescreen of supercapitalism that hides the tyrannical intent of the mega-industrialists who are seeking to monopolize control of all key industries in order to rule the entire earth.

One way they accomplish that is through the nationalization of industries. As long as these industries are in control of the government, they are, in essence, under their rule. The danger we face in our country is the movement toward the socialization of our nation. In the end, that will bring with it the nationalization of the major industries, completely destroying the free enterprise system that brought our nation to such prosperity. Remember that the enemy of freedom and liberty is, was, and always will be the centralization of power and the economy.

Notes

1. David M. Kennedy, *The American People in the Great Depression: Freedom from Fear: Part 1* (Oxford, UK: Oxford University Press, 1999), 125.
2. Brian Hodges, "Breaking: Government Conspired to Deprive Nevada Rancher of Property Rights," May 29, 2013, Pacific Legal Foundation, https://pacificlegal.org/breaking-government-con-spired-to-deprive-nevada-rancher-of-property-rights/.
3. Ravi Zacharias, *The End of Reason: A Response to the New Atheists* (Grand Rapids, MI: Zondervan, 2008), 122.
4. Ibid, 126.
5. Stuart Schram, ed., *Chairman Mao Talks to the People*, trans. John Chinnery and Tieyum (New York: Pantheon Books, 1974), 82.
6. John Stossel, "Sweden Isn't Socialist," January 2, 2019, Capitalism Magazine, https://www.capitalismmagazine.com/2019/01/sweden-isnt-socialist/.
7. Stuart Schram, ed., *Chairman Mao Talks to the People*, 111.
8. Nigel Cawthorne, *The Crimes of Stalin: The Murderous Career of the Red Tsar* (London: Arcturus Publishing, 2011), 106.
9. Ibid., 114.
10. Ibid., 117–118.

CHAPTER 7

● ● ●

THE RISE OF SUPERCAPITALISM IN AMERICA

Supercapitalism is the unjust distribution of the equality of opportunity, created by the rich, that unjustly destroys all competition and establishes giant monopolies in order to hoard control of wealth and power. We have seen this developing in Europe for some time, and now it is also congealing in the United States. How did we get there?

In the first century after the birth of the American republic, the Industrial Revolution brought enormous growth and wealth to our fledgling nation. However, with that growth came new dangers to the American experiment of self-government. The meteoric rise in wealth for a few industrialists brought enormous financial power to an elite few who sought to unjustly monopolize the more important sectors of our economy. These sectors include but are not limited to banking, insurance, energy, transportation, communications, and military defense.

In 1879, C. T. Dodd, attorney for the Standard Oil Company of Ohio, devised a new type of trust that allowed Standard Oil to monopolize the oil industry by being able to own the stocks of competing companies through a board of trustees that maintained the operational controls over several companies.

Public sentiment against this threat to our nation's ideals grew steadily until the pressure forced politicians into action. On July 2, 1890, Congress passed the Sherman Antitrust Act, and President Benjamin Harris signed it into law. The act states:

> Every contract, combination in the form of trust or otherwise, or conspiracy, in restraint of trade or commerce among the several States, or with foreign nations, is declared to be illegal. . . . Every person who shall monopolize, or attempt to monopolize, or combine or conspire with any other person or persons, to monopolize any part of the trade or commerce among the several States, or with foreign nations, shall be deemed guilty of a felony.[1]

The act, however, is brief and unspecific, giving a great deal of latitude to the U.S. Supreme Court in its interpretation. But contrary to public misconception, the act was not intended to prevent the dominance of an industry by a specific company. According to the act, a monopoly achieved by merit is considered perfectly legal.

In other words, the act simply made illegal the artificial raising of prices by the restriction of trade or supplies. Any monopolies built innocently were perfectly legal. This enormous legal loophole has grown wider with time as the courts have made a clear distinction between coercive and innocent monopolies.

Modern courts have progressively come to hold plaintiffs to a much more stringent burden of proof of a conspiracy that would label a defendant a coercive monopoly. Moreover, it has resulted in the protection of defendants from bearing the cost of antitrust fishing expeditions that deprive the plaintiffs of, perhaps, their only real tool to acquire evidence from the internal information in these giant institutions that can outlast any prosecutor with their well-financed legal teams.

That has resulted in the quick resolution of most cases in favor of the monopolies before any discovery is even done. Critics of the weak law have charged that it was redacted as a smokescreen to allow monopoly protectionists the latitude of later passing a more important tariff law that allowed big business to further bilk the public.

Economics professor Thomas DiLorenzo has noted that Senator John Sherman, who introduced the Sherman Antitrust Act, also sponsored the 1890 tariff (a pro-trust law relating to tariffs) just three months after his antitrust act was passed.

> Protectionists did not want prices paid by consumers to fall. But they also understood that to gain political support for high tariffs they would have to assure the public that industries would not combine to increase prices to politically prohibitive levels. Support of both an antitrust law and tariff hikes would maintain high prices while avoiding the more obvious bilking of consumers.[2]

Moreover, the Sherman Antitrust Act labeled labor unions as cartels and provided a legal tool to outlaw and legally disband labor unions that fought to overcome the many injustices of big business. This glaring inequity was later addressed in the Clayton Antitrust Act of 1914 (Section 6) as public sentiment against the abuses of big business grew to a fevered pitch.

The Clayton Antitrust Act also outlawed mergers and acquisitions where the effect may substantially lessen competition. However, the conduct is only considered illegal and the plaintiff can only prevail if it is proved to the court that the defendants are doing substantial economic harm. This burden of proof has also become ever more elusive for prosecutors.

However, there is a dark side to those who, in the name of antitrust laws, can confiscate private property that has been honestly earned to parcel it out to others. This does not serve the free market form of capitalism and opens up the avenue for politicians to illicitly parcel out to others what has been honestly gained. So there are two edges to this sword, and justice must walk a tight rope that will not destroy the free market system.

> Frequently, when government invokes the antitrust laws, it transforms a company's private property into something that effectively belongs to the public, to be designed by government officials and sold on terms congenial to rivals who are bent on the market leader's demise. Some advocates of the free market endorse that process, despite the destructive implications of stripping private property of its protection against confiscation. If new technology is to be declared public property, future technology will not materialize. If technology is to be proprietary, then it must not be expropriated. Once expropriation becomes the remedy of choice, the goose is unlikely to continue laying golden eggs.
>
> The principles are these: No one other than the owner has a right to the technology he created. Consumers can't demand that a product be provided at a specified price or with specified features. Competitors are not entitled to share in the product's advantages. By demanding that one company's creation be exploited for the benefit of competitors, or even consumers, government is flouting core principles of free markets and individual liberty.[3]

The key phrase in my opinion is quite important for us to keep in balance: "If new technology is to be declared public property, future technology will not materialize. If technology is to be proprietary, then it must not be expropriated. Once expropriation becomes the remedy of choice, the goose is unlikely to continue laying golden eggs."[4] Finding that balance between a government that secures the right of private property and the deterrent of those who would seek to monopolize industry must be our course if free market is to be real and justice is to be served. It is at this juncture that progressivism wields its weapons of confiscation and completely eradicates our right to private property.

Nevertheless, it is my sincere opinion that our Constitution had it right when it gave the representatives of the people the right to control our national finances and our currency. Today, the Federal Reserve, which is not a government agency but a cartel of huge international bankers that lends our government money to do its business, is the entity that has the real power to regulate our currency.

Is it any surprise that our government has, since the time of Andrew Jackson, operated in a deficit that has been steadily skyrocketing? Every dollar that our government runs in deficit is owed to the international banking cartel that charges interest on the debt and makes a killing on our hopelessly unbalanced budget. Guess who pays for that debt? You and I and our children and grandchildren! The wolf is guarding the henhouse. How did it come to that?

The Rise of Progressivism in Academia and Christianity

With the meteoric rise of robber barons toward the end of the nineteenth century and the social injustices created by their overwhelming greed, the stage was set for the rise of progressivism in America.

By the end of the nineteenth century, the Enlightenment ideology had rooted deeply into the academic circles of Europe, and it began to influence the academic institutions in America. In Russia and throughout most of Europe, the Marxist ideology grew in popularity as Christianity's influence ebbed considerably.

Here in America, Baltimore's Johns Hopkins University was founded in 1876 as a new type of academic research institution that promoted the Enlightenment ideology. It sought to replicate the

Prussian academic tradition of Heidelberg, Freiberg, Gottingen, and Berlin. The Marxist-Hegelian socialist ideology began to infiltrate into American academics and challenge the classical Judeo-Christian precepts of individual liberties ordained by God, as well as the deep-rooted American idea of a small federal government and the decentralization of power, which were seen as absolutely necessary due to the fallenness of humanity. The ideology of our Founding Fathers was rejected in favor of an expanded state and a collectivist ideology that saw humans not as fallen creatures but as those who could perfect both society and individuals through their own self-efforts.

The Nine Core Principles of Progressivism

1. Deny the Judeo-Christian principle of the fallenness of humanity.
2. Both the individual and collective society are perfectible through the auspices of a strong federal government that can impose the necessary force to make its citizens comply.
3. Belief in the centralization of power and the diminution of local power.
4. The Constitution should be filtered through the Darwinian ideology in order to make the arbitrary changes necessary to foster the idea of a perfect society.
5. The ends justify the means, and all pragmatic choices to accomplish that means are morally acceptable; morals are relativistic.
6. The need of the collective outweighs the needs of the individual; thus individual freedom must be sacrificed for the good of the collective.
7. Belief in forced redistribution of wealth to create a more just society with a goal to raise taxes for the working people in order to provide social assistance to those who do not work.

8. Have not yet declared that they do not believe in the right to own property (the only difference between progressivism and communism). But the state has the right to confiscate private property for whatever purposes it deems a prerogative, as seen in the health industry in the takeover of private insurance companies and all hospitals and medical practices, and the takeover of energy as seen in the attempt under the guise of global warming to control all fossil fuels.

9. Expert and enlightened knowledge of social issues justifies their right to impose their will on the unenlightened common people, and they have thus taken over all public educational institutions in order to indoctrinate America's youth in their socialist agenda.

At the beginning of the twentieth century, inroads of Enlightenment ideology into almost all major evangelical Christian denominations began to change the traditional evangelical worldview into a liberalized neo-orthodoxy. With the rise of higher criticism, many evangelical institutions changed drastically from their traditional evangelical doctrines and began to teach a social gospel in which the main emphasis of their missionary work was couched in Marxist ideology that favored the rights of the collective over the rights of the individual.

As progressivism infiltrated Christian denominations, it led to numerous splits in evangelical denominations—a process that continues even today. The concern of these progressivist Christians changed from the traditional Christian message of the personal salvation of the individual through the work of Christ on the cross to a message of perfecting society and humans by perfecting both government and individuals through a coercive socialist ideology. Beck further states,

The most prominent of the first American progressives was Richard T. Ely, a professor of economics who came to Johns Hopkins in 1881, two years after receiving his doctorate at

Heidelberg. Ely, who once wrote that "God works through the State in carrying out his purposes more universally than through any other institution," helped found the American Economic Association, which is dedicated to social science and social justice (and which still holds an annual lecture named for him).[5]

Ely promoted the socialist ideology of forced redistribution of wealth. Although he did not reject the right of private property that is central to the collectivist ideology, he favored the forced redistribution of the profits of all business enterprises:

> Over time, Ely trained hundreds of social scientists in progressivism and his views about the "perfectibility" of society and man, but two of his disciples stood out: Woodrow Wilson and John Dewey.
>
> When they first became attracted to Ely's ideas, Dewey was teaching high school and Wilson was working as a lawyer. . . . Dewey went on to become a career academic and progressive educational reformer, arguing that only a far larger governmental apparatus could cure the social ills of the twentieth century. He argued that freedom was not "something that individuals have as a ready-made possession"; it was "something to be achieved." In this view, freedom was not a gift from God or nature; it was a product of human making, a gift from the state. He emphasized state influence on early-childhood education in order to spread the progressive doctrine to children as early as possible, no matter what views they were exposed to at home.
>
> Progressive academics such as Dewey and Wilson, who eventually left the legal profession to teach (first at Cornell and Bryn Mawr and then at Princeton), had an

ally in American Protestantism. Most social scientists such as Ely and Wilson were devout Christians themselves and open about their desire as Christian missionaries to build a kingdom of heaven on earth. This was the "social gospel," a vision of Hegel's and Ely's progressivism that sought economic and social improvement by applying Christian ethics.[6]

I would only like to add one important correction to Glenn Beck's excellent statements: These were not devout Christians, even though they went by that title. They did not believe in the traditional Christian gospel but simply dressed their collectivist ideology in Christian nomenclature. The early progressives understood well the deeply entrenched American Christian values and cleverly disguised their collectivist ideology in Christian garb to make it more amenable to the American mind. This tactic was quite clever and successful.

In fact, it was for that same reason that Ely pragmatically understood that he would gain no footing in America if he attacked the right to own private property. It was another clever smokescreen that hoodwinked many in America. Ever does the Enemy of Man seek to undermine and corrupt the truth with clever smokescreens that hide his true intent. But make no mistake, the confiscation of all private property has always been the final goal of all collectivist ideologies.

From the beginning of our nation, the distinctly American spirit of individualism and distrust of the power of governments to curtail our individual freedoms ruled the hearts of all Americans. But a new threat to these freedoms blossomed with the enormous leap in industrialization after the Civil War. With the meteoric rise of robber barons at the end of the nineteenth century and the social injustices created by their overwhelming greed, the stage was set for the rise of progressivism in America, which promised to bring a more just economic condition for the working person.

American workers feared that these elite industrialists would become the new nobility that would once again plunge them into the tyranny they had so bravely fought against in the American Revolution. Progressivism began to make inroads into the unions and to gain support from politicians in both the Republican Party and the Democratic Party.

From the beginning of Thomas Jefferson's tenure as president in 1801 to Grover Cleveland's last year as president in 1897, the Democratic Party stood firmly against a strong federal government. Their opposition ended that year. Progressives claimed that the only way to control these powerful industrialist elites was through the power of a strong and expanded federal government that could protect the people; it was the same argument that was being promoted by Vladimir Lenin in Europe and Russia.

But the Democrats were not alone in this regard. Republicans also began pushing the progressivist agenda. In fact, the first progressive president was Theodore Roosevelt, a Republican, who began his tenure in 1901. He was the first president of the United States to endorse a federal income tax and a national health insurance program. He expanded the power of the executive branch as none other before him. He bought large tracts of land so the federal government could establish national parks. He superficially attempted to gain control of the supercapitalist monopolies by expanding the power of the executive branch.

> In 1910, Roosevelt declared, "The absence of effective State, and, especially, national, restraint upon unfair money-getting has tended to create a small class of enormously wealthy and economically powerful men, whose chief object is to hold and increase their power." He also said that the government "should permit it [their fortunes] to be gained only so long as the gaining represents benefit to the community."

This was a Republican essentially saying that private wealth is only allowable to the extent that it benefits the greater good. Roosevelt also argued that accumulated property is "subject to the general right of the community to regulate its use to whatever degree the public welfare may require it." He advocated concentrating power in the Presidency to make this system work. "This New Nationalism," he said, "regards the executive power as the steward of the public welfare."[7]

This new nationalism was the beginning of the establishment of the progressivist agenda in our nation, a process that continued with few exceptions all the way to President Obama. In his first inaugural speech, Obama called his ideology a "fundamental transformation." Few suspected then that this fundamental transformation meant the change from a capitalist economy to a socialist economy. As of 2019, the Democratic Party has a considerable number of radical leftist socialists who are hell-bent on changing the United States from a capitalist nation to a socialist nation in order to eventually merge us into a global charter.

At the beginning of the twentieth century, three issues that would determine the future economic stability of the American middle class were paramount in the hearts of the American people: (1) the gold standard, (2) the danger of giant monopolies to free enterprise, and (3) the debate over the control of the national currency.

The founders of the U.S. Constitution specifically stipulated that Congress is the sole entity with the right to control fiscal matters and ought to be the only one responsible for the issuance of our national currency. But pressure from big business to establish a central national bank was mounting with great force. The industrialists were attacking the stability of our economy as a ploy to force upon the public the establishment of a national bank in order to supposedly assure economic stability.

Another important factor in the stability of the currency was the issue of the gold standard. Historically, the currency issued by the American government was fixed in terms of a specified amount of gold—called the gold standard.

In other words, paper money could be freely converted into gold; every dollar in paper money was backed by a dollar's worth of gold. Most of the time prior to the twentieth century, America had a bimetallic (gold and silver) system of money, which was essentially a gold standard since very little silver was traded. In 1900, however, America turned to a true gold standard with the passage of the Gold Standard Act.

This is enormously important to the middle class, which has no great reserves of wealth because that keeps their hard-earned money from greatly depreciating. If the savings accrued for retirement in most American middle-class families is greatly devalued, then they are incapable of meeting their needs during their retirement years. This heartless ruse is tantamount to the outright theft of their life's labor, leaving them destitute at a time they can no longer work.

At the turn of the twentieth century, most Americans were well aware of this issue. Today, only 100 years later, the vast majority of Americans are completely unaware of the precarious financial position into which they have been led. The hard-earned savings of the Baby Boomers, who are now entering retirement, could go up in smoke in a single day.

The gold standard effectively came to an end in 1933 when President Franklin D. Roosevelt outlawed the private ownership of gold. The government then passed the Bretton Woods system (1946), which created a modified form called the gold bullion standard.

Under that system, the government holds bars of gold (bullion) to back the token currency. It provides a fixed system of exchange rates that allows foreign governments to sell their gold to the United States Treasury at $35 per ounce. Private ownership of gold is outlawed, but the government may trade in gold. That at least provided a formal link between the world currencies and gold, which gave some measure of stability to the American dollar.

Since the supply of gold limits the amount of currency in circulation, it inhibits the government from the temptation to pay its debts by simply printing more money, which causes inflation and devalues the currency. If a nation buys more abroad than it sells, then it must export gold to pay its debt. That reduces the amount of gold and thus the available currency in our nation, which then causes the prices of goods to fall accordingly.

Unfortunately, once the absolute gold standard was removed, the value of the dollar in relation to the rate of the amount of gold held in reserve could be altered so the fixed rate is no longer an absolute standard. In 1972, the American dollar was again arbitrarily devalued from 1/35 of a troy ounce of gold to 1/38. Then again in 1973, it was further devalued to 1/42 of a troy ounce of gold.

Until March of 1968, our currency was at least backed by 25 percent gold reserves. That, however, came to an abrupt end on August 15, 1971, when President Richard Nixon disbanded the Bretton Woods system, breaking for the first time in modern human history the formal link between world currencies and gold. This was an unmitigated disaster for the middle class, and it must be corrected.

Almost every major country is now in a financial system of fiat money. In other words, money has no real value other than the faith of its users in its ability to be a means of exchange for goods. It is intrinsically worthless paper that cannot be traded for a fixed amount of gold and that can be greatly devalued through the fiat decision of the Federal Reserve to simply print more money. We have given the wolves of Wall Street the keys to the henhouse, and if we do not act to change it, we will pay a heavy price in the not too distant future.

The Wolf Guarding the Henhouse

Our national finances are today squarely in the hands of the Federal Reserve bankers. This is an enormous financial power that can be used by big business to whittle away the buying power of the middle class by the devaluation of their hard-earned savings.

It would be a little less dangerous if Congress still had power over our currency. If the government is responsible to the people, and since the people have power over their elected officials by their votes, then there is at least a checks and balance system in play to oversee the whole affair. Of course, the problem with this is that politicians can and often have been bribed by the elite in power. That is how they managed to institute the Federal Reserve system to begin with. It nevertheless creates a buffer between the outright manipulation of finances by the bankers and the opportunity for our votes to counter their agenda.

But what is truly alarming is the fact that in direct opposition to the mandate of our Constitution, Congress is no longer the power that regulates our currency. According to Article 1, Section 8 of the U.S. Constitution, the founders of the United States of America delegated solely to Congress the power to, among other things, "coin money, regulate the value thereof, and of foreign coin, and fix the standard of weights and measures."[8]

Congress has abrogated its constitutional authority and turned it over to big business through the vestiges of the Federal Reserve. **The wolf is now guarding the henhouse, and there are no guards that oversee the wolf. They do not even require a congressional audit to hold them accountable. Can you imagine that?**

All the well founded fears the general public had at the turn of the twentieth century regarding a centralized bank have been realized. Thomas Jefferson and Andrew Jackson are turning in their graves. Unfortunately, most of us have been cleverly hoodwinked into ignorance, inaction, and abject apathy in these matters so crucial to our national and individual financial security.

The antitrust measures that were passed at the end of the 1800s at least partially sought to create a more just system that ensured the freedom of the small investor, or what we refer to in our American colloquialism as the mom and pop stores or businesses. These anti-trust laws also provided a measure of protection for the consumer, but all of it has now been effectively circumvented.

Since even before the presidency of Andrew Jackson, there was an unrelenting and concerted drive by a few elite industrialists to gain control of our government's economic power. These financial tycoons, during the greater part of the nineteenth century, initially sought to make Congress abrogate federal powers endowed wisely by our Constitution in regard to the printing of money and the financial affairs of our nation, and actively crusaded to turn control over to a private syndicate of international bankers. Their scheme had existed for decades before they succeeded with Woodrow Wilson.

During Andrew Jackson's tenure as president, he fought bravely to keep bankers from establishing a central bank to take over the American economy. An assassin at point blank range fired two pistols at his stomach, but by God's grace, both pistols misfired. Jackson successfully kept industrialist bankers at bay during his tenure of office, an act that is memorialized in his tombstone by his own decree.

Prior to Jackson, as a result of the economic chaos from the War of 1812, Congress during James Madison's presidency passed a law creating the Second Bank of the United States. It was given a 20-year charter that went into effect in 1816. The so-called Bank War began when Nicholas Biddle, the bank's president, and his ally Henry Clay, perhaps one of the most powerful political figures of his day, attempted to renew the charter in 1832. When Jackson defeated Clay for the presidency, he promptly vetoed the renewal of the bill and effectively killed the chartering of the central bank.

> The veto held up when Jackson defeated Clay in the 1832 presidential election. Biddle did not go quietly, however. When Jackson began transferring the federal government deposits out of the Second Bank to his favored "'pet banks,"

the Second Bank demanded payments on bills issued by state banks and reduced its loan by over $5 million, contracting the money supply and causing interest rates to double to 12 percent. Biddle hoped, by damaging the economy, to stir up opposition to Jackson; in the process, he showed that Jackson had not been wrong to fear the power of a major bank to distort the economy for its own purposes.[9]

That power to extort the government through financial chaos that can be artificially created by the financial elite became a very worrisome problem for the common citizen. It would rear its head again three more times before our present time. But in the meantime, Jackson had quashed the centralization of the American banking system.

Most important, Jackson's victory ensured that a powerful private bank was not able to install itself in the corridors of political power and use its privileged position to extract profits for itself, inhibit competition, and hamper broader economic development.[10]

Popular concern for this danger from big business grew steadily in our maturing nation. The rise of the robber barons generated a deep concern among the common citizen that the economic power of these elite industrialists was becoming quite threatening to the majority of individuals who, like Jefferson, feared an economic takeover of our political system by a few wealthy elite industrialists.

A new nobility was rising in America that threatened the vision of our Founding Fathers as a government that is responsible to all the people; in other words, a government that governs by the consent of all its citizens and not a shadow government of elite financial tycoons who control politicians through their great wealth.

The innovations that changed the economic landscape changed the political landscape as well. Social mobility and the lack of an entrenched aristocracy meant that newly successful companies and industries could gain political representation quickly, at least when compared to European societies. New money could make its way into politics, whether legally or illegally. By the late nineteenth century, the Senate had become known as the "Millionaires' Club"; buying political support with cash was considered by many to be just an extension of normal business practices.[11]

Railroads were built to connect our nation with commerce, and enormous fortunes were made when people could travel much farther in shorter periods of time to bring goods from either side of the United States and compete in the market. With those enormous fortunes came a growing political power.

The railroad barons and their industrial allies acquired great political power, coming to dominate the Senate by the turn of the century.[12]

The stage was being set once again for an attempt to control the financial power of the nation. The vast majority of Americans was quite disturbed at this possibility. Jackson had previously fought against paper currency and preferred hard money—gold and silver. He firmly believed that paper money would allow bankers to distort the economy at the expense of the common people.

Most Americans at that time also held that view, but in the end, paper money won the day. Nevertheless, as a safeguard against this manipulation of paper money by financiers, the people demanded that paper money be backed by gold.

In 1897, President William McKinley acknowledged this public mandate in his first inaugural address, having handily won the election primarily on the issue of the gold standard and the retention of the power of Congress over the issuance of the national currency as well as the reduction of the expenditures in the federal government. The American public, by their votes, was able to circumvent the will of the financial elite.

McKinley's inaugural address is as relevant to the concerns of our nation today as it was on the day he gave it. Sadly, our people are nowhere near the level of concern or knowledge of this issue. Just 100 years ago, the politically correct and constitutionally incorrect modern view that government and God ought to be permanently divorced did not exist.

In obedience to the will of the people, and in their presence, by the authority vested in me by this oath, I assume the arduous and responsible duties of President of the United States, relying upon the support of my countrymen and invoking the guidance of Almighty God. Our faith teaches that there is no safer reliance than upon the God of our fathers, who has so singularly favored the American people in every national trial, and who will not forsake us so long as we obey His commandments and walk humbly in His footsteps. . . .

The country is suffering from industrial disturbances from which speedy relief must be had. **Our financial system needs some revision; our money is all good now, but its value must not further be threatened. It should all be put upon an enduring basis, not subject to easy attack, nor its stability to doubt or dispute. Our currency should continue under the supervision of the Government.** The several forms of our paper money offer, in my judgment, a constant embarrassment of the Government and a safe balance in the Treasury. Therefore I believe it necessary to devise a system which, without diminishing the circulating medium or offering a premium for its contraction, will present a remedy for those arrangements which, temporary in their nature, might well in the years of our prosperity have been displaced by wiser provisions. . . .

The severest economy must be observed in all public expenditures, and extravagance stopped wherever it is found, and prevented wherever in the future it may be developed. **If the revenues are to remain as now, the only relief that can come must be from decreased expenditures. But the present must not become the permanent condition of the Government. It has been our uniform practice to retire, not increase our outstanding obligations, and this policy must be resumed and vigorously enforced.** Our revenues should always be large enough to meet with ease and promptness not only our current needs and the principal and interest of the public debt, but to make proper and liberal provision for that most deserving body of public creditors, the soldiers and sailors and the widows and orphans who are the pensioners of the United States. . . .

It has been the policy of the United States since the foundation of the Government to cultivate relations of peace and amity with all the nations of the world, and this accords with my conception of our duty now. **We have cherished the policy of non-interference with affairs of foreign governments wisely inaugurated by Washington, keeping ourselves free from entanglement, either as allies or foes, content to leave undisturbed with them the settlement of their own domestic concerns** (emphasis added).[13]

McKinley's platform called for the establishment of the gold standard in our currency. He insisted that the responsibility of the national currency should remain under the auspices of Congress and that our fiscal policies should be to maintain a balanced budget. He also resisted the internationalist-globalist policy, citing George Washington's wise recommendation in that regard.

All four of these fundamental elements held dear by the common folk were in direct contradiction to the will of big business that had long been campaigning to make a central national bank the entity in charge of our currency. It stood to make a tremendous amount in interest on loans to pay for accrued public debt.

The international financiers also sought a globalist agenda for their economic imperialist purposes. Sadly, all four of those goals have been accomplished in our nation, contrary to the welfare of our common folk and the wishes of our Founding Fathers. We have fallen from a great height, and few are even aware of it, so successful is the indoctrination of our public educational system that was largely influenced by Dewey's progressivist agenda.

Sadder yet is the fact that most Americans have absolutely no idea of the danger it creates for their financial stability or the clever tactics that were used to accomplish these nefarious ends. At any moment, a financial breakdown can be artificially engineered that will totally throw our nation into a maelstrom and undermine our savings and financial power, which will be unable to offer any concerted resistance.

Like Andrew Jackson before him, William McKinley became the enemy of those who wanted to control our American financial system through a centralized banking consortium. On September 6, 1901, after McKinley won his second term of office and while he was attending a function at the Pan-American Fair in Buffalo, New York, an assassin shot McKinley, firing two rounds at point blank range. The first bullet gave him only an insignificant flesh wound in the shoulder, but the second bullet lodged in his stomach. The doctors did not operate to remove the bullet, claiming that it might do more harm to do so.

McKinley was a devout Christian and a kind, loving husband whose humility and integrity caused all who knew him to love him.

318

He broke presidential protocol many times to be by his wife who had been gravely ill.

Such was McKinley's character that after being shot, while his bodyguards proceeded to pummel the assassin, he leaned over and said, "Go easy on him, boys." Several days after being shot, McKinley died of gangrene from his wounds. His last words to his wife who was at his bedside were, "It's God's way. His will, not ours, be done."[14]

Theodore Roosevelt, the vice president, became president after McKinley, succumbed to his wounds. The lone gunman was said to have been an anarchist, but although the gunman had met some prominent anarchists, he had no formal ties to any. That charge is a bit dubious to me.

Roosevelt's first wife was Alice Hathaway Lee, the daughter of a prominent New England banking family. Unfortunately, she died of typhoid fever after giving birth to their daughter, Alice. Roosevelt later remarried the refined Edith K. Carow, a friend from his childhood.

In 1900, the prominent political boss Thomas Platt had urged McKinley to take Roosevelt as his running mate. Roosevelt had initially resisted but later agreed to run for vice president.

Although he has popularly become known as a trust-busting president, Roosevelt's true inclinations were actually to regulate giant industrialists and not do away with them altogether. Nevertheless, his efforts in forcing arbitration upon management in the coal strike of 1902 and his use of the Sherman Act against such monopolies as Standard Oil Company of Ohio caused great consternation to big business.

Roosevelt asked Congress to create a Department of Commerce and Labor and a Bureau of Corporations that were authorized to investigate business combinations. He brought suit against Northern Securities that had merged with Northern Pacific, the Great Northern, and the Burlington Systems, successfully bringing them to court and causing their dissolution. Suits were also filed against United States Steel Corporation and some railroad monopolies during Roosevelt's tenure in office.

Under his leadership, the conservation movement made great strides. But his antitrust policies brought him into disfavor with the interests of big business. Wealthy industrialists, seeing the increasing shift in momentum of the public toward antitrust laws, decided to take action in order to quash the burgeoning danger.

It has become generally accepted by most historians that J. P. Morgan and his international financier colleagues artificially created the depression of 1907. The strong-arm tactic of Nicholas Biddle in the previous century was refined and executed by J. P. Morgan. A battle for the control of banks had previously ensued between F. Augustus Heinze of Knickerbocker Trust Company and J. P. Morgan, who mounted an attack on Heinze's trust company, which negatively affected the American middle class in its wake, causing thousands of businesses and smaller banks to fail.

Heinze was forced to resign as bank president, and on October 21, 1907, the National Bank of Commerce ceased to honor checks drawn by the Knickerbocker Trust. That caused a run on the trust, and by the next day, the Bank of North America had failed. The panic created by the run on the Knickerbocker Trust further incited runs on nearly all the trust banks, causing them to also fail. As a result, many of J. P. Morgan's rivals and their cronies were effectively run out of business.

Then, in order to look good in the public eye, J. P. Morgan came in to save the day and stop the escalating depression that was created by their strong-arm tactics. He organized a group of his international financiers to raise up all the failing companies and companies whose stocks were plummeting due to the artificially created economic instability.

The group injected money into the economy by redirecting money to these failing banks, which they took over for pennies on the dollar. In one fell swoop, the industrialists flexed their economic muscles and gave notice to the government of their unrivaled power to wreck the American economy. At the same time, they gained unrivaled control of many competing banks in the East.

The Panic of 1907 was then used as an excuse to ramrod the formation of a central bank to supposedly avert similar problems. It is my opinion that this was simply nothing less than outright extortion. In just six years, the Federal Reserve was given the responsibility of supervising our currency, a right that rightfully belongs to the representatives of the American people in Congress by constitutional mandate.

Just prior to that, after his second term, Theodore Roosevelt had announced that he would not seek reelection, following the example of George Washington who could have been reelected to a third term. Nevertheless, when Roosevelt realized that the two candidates vying for the presidency threatened to undo the work he had accomplished during his tenure, he decided to begin a new party—the Progressive Party, which he called the Bull Moose Party—and enter the presidential race. Believe it or not, the Progressive Party began as a fracture of the Republican Party. It is fair to say that both Republicans and Democrats have been financial puppets of the oligarchy for more than 100 years.

Roosevelt once again entered the fray with his usual inexhaustible energy and bravado, gaining great popularity; that is, until an assassin attempted to take his life during one of the speaking engagements of his whirlwind campaign. He was shot in the chest at point blank range but continued giving his speech until he was finished. Then he was taken to the hospital.

He reasoned that the bullet had not reached his lungs since he was not coughing up blood, and he therefore insisted on finishing his speech. The 32-caliber bullet had actually hit his metal eyeglass case and a thick layer of paper, his rolled-up speech that he kept in his coat pocket. Both of them miraculously saved his life.

However, I find it quite curious and in disaccord with his general demeanor that Roosevelt quit campaigning after he was shot. His retreat is highly suspicious to me and quite out of character. We may never know this side of heaven what the outcome of that election would have been had he continued to campaign.

This split in the Republican Party was then blamed for giving the election to the Democrats, and Woodrow Wilson became the next president. That single 32-caliber bullet that hit Roosevelt may have caused our nation the most important concession to big business that sealed its eventual takeover of our national economy.

The Great American Betrayal

Woodrow Wilson was also a progressivist but in character nothing like Theodore Roosevelt, the rugged outdoorsman, the people's man. Wilson was an acidly elitist intellectual whose arrogance was only superseded by his narcissism. He sincerely believed that his expert knowledge gave him the right to dictate and control public policies for common and unenlightened people.

> Wilson didn't spend decades in academia simply to learn. To him, higher education was a tool to hone a new philosophy of American government led by Hegelian experts focusing on the collective instead of the individual, an elite cadre of intellectuals at the helm working to perfect society. Wilson plotted to be the captain of that ship.
>
> His time at Johns Hopkins only helped cement his big-government attitudes. Many of that university's early professors were German-trained. Through them—particularly the influential early progressive economist Richard Ely—Wilson lapped up an admiration for Prince Otto von Bismarck and the powerful new authoritarian German welfare state. He also imbibed a belief in Darwinism, concluding that a more powerful, centralized government was critical to society's evolution.
>
> It all added up to an absolute infatuation with governmental power. "If any trait bubbles up in all one reads about Wilson," observed historian Walter McDougall, "it is this: he loved, craved, and in a sense glorified power."[15]

But for all of his supposed intellectual prowess, he was outfoxed by the robber barons when they convinced him to support the establishment of a central bank. No doubt, the idea of a centralized bank, as proposed through the Federal Reserve, played into his idea of a more powerful executive branch and the centralization of power that obsessed him. After all, the plan called for the president to select who would chair this centralized banking system.

That power, which, according to the U.S. Constitution, was to reside in the people's representatives in Congress, would now be somewhat under the wing of the executive branch. But this was nothing more than an illusion. The president has no real authority over the Federal Reserve, and neither does any other part of our government.

In the meantime, Senator Nelson Aldrich (Rhode Island), chair of the National Monetary Commission that was established to propose legislation that regulated banking, was busy behind the scenes pushing for the establishment of the Federal Reserve to take over the responsibility of our national currency. Aldrich was a wealthy Rhode Island merchant and a public utilities lord. His daughter married John D. Rockefeller, Jr. His son Winthrop W. Aldrich later became chair of the Rockefeller-controlled Chase National Bank.

In November 1910, a group of these international financiers met secretly on Jekyll Island, Georgia, to hammer out the details of this nefarious enterprise. On the evening of November 22, 1910, Senator Aldrich met with A.P. Andrews, Assistant Secretary of the Treasury Department; Paul Warburg, a naturalized German representing Kuhn, Loeb & Co. who later provided arms for the West in World War I and whose brother Max Warburg did the same for Germany, and who was an advisor to Kaiser Wilhelm II; Frank A. Vanderlip, president of National City Bank of New York; Benjamin Strong, representing J.P. Morgan Company; and Charles D. Norton, president of the First National Bank of New York, also controlled by Morgan, and who left Hoboken, New Jersey, in a private train bound for Jekyll Island.

Kuhn Loeb and Company was initially founded by Abraham Kuhn and Solomon Loeb as a mercantile company in the mid-nineteenth century, but upon the marriage of Jacob Schiff and Solomon Loeb's daughter Theresa, the company set up in Wall Street as a private bank. Paul Warburg, who came from a German banking house, married Nina J. Loeb, and thus the Warburg connection with Kuhn and Loeb. Felix M. Warburg, also Paul's brother, married Frieda Schiff, and thus the Warburg-Kuhn-Loeb-Schiff-Kahn financial dynasty was established. One of Jacob Schiff's descendants is Andrew Newman Schiff, who is married to former Vice President Al Gore's daughter Karenna.

This group of men gathered on Jekyll Island represented one-sixth of the entire wealth of the planet. There, in the secret seclusion of that island, plans were set in motion to establish the Federal Reserve system. The Aldrich Plan was set up and presented to Congress.

However, the Aldrich plan was politically controversial; it looked like a trick to get taxpayers to finance banks and protect them from the consequences of their risky activities. Opponents argued that the problem was a cabal of big banks that were secretly running the country. . . . (The investigation was proposed by Representative Charles Lindbergh Sr., who called the Aldrich plan a "wonderfully devised plan specifically fitted for Wall Street securing control of the world.") The Pujo Committee concluded that control of credit was concentrated in the hands of a small group of Wall Street bankers, who had used their central place in the financial system to amass considerable economic power. The committee report provided ammunition to Louis Brandeis, a prominent lawyer and future Supreme Court justice. . . . Brandeis spoke out strongly in favor of constraining banks.[16]

Just four years after that secret meeting on Jekyll Island, Wilson ignorantly signed the Aldrich Plan into law. In 1914, the second progressive president provided not for the protection of the people from the robber barons but instead made the robber barons the direct profiteers of big government. The wily robber barons realized they could not stop the rising tide of public indignation of their political power due to their enormous wealth, so they decided it would be better for them to embrace the progressivist movement rather than fight it with assassins and bullets. That single treasonous signature by Woodrow Wilson sealed the rule of the oligarchy in America. It was the day of the great American betrayal.

Fears regarding a Wall Street takeover of our financial system as an insurance policy that protected them with tax-dollar bailouts for their risky and slimy financial concoctions were proved correct in 1929 and again in 2008 when the government provided monumental bailouts at taxpayers' expense.

The flaw in the rationalization for forming the Federal Reserve was to stem an economic crash. The most lucrative moments for international financiers are during an economic crash. They get to deal in risky activity, knowing full well that they have a golden parachute—the uninformed American taxpayer. And when small companies fail due to the economic upheaval, they buy them up for pennies on the dollar, further consolidating their financial hold on the American economy.

It took only sixteen years for this flaw in the system to become catastrophically apparent. Rampant speculation in the 1920s led to the Crash of 1929, which was initially followed by a generous bailout for elite New York financial firms, and then by the repeated bungling of attempts to save the rest of the financial system. Not only did the Federal Reserve's safety net encourage excessive risk-taking by bankers; the safety net, it turned out, had gaping holes that could not be fixed in the intense pressure of a crisis. The result was the Great Depression.[17]

The failed companies sold anything they could at whatever price they could get to offset their losses and pay off at least some of their debts. Prices plummeted in a downward spiral, and millions of Americans became bankrupt overnight. It is hard for us to imagine how the hard-earned savings of a person's entire lifetime, intended to provide a financial safety net, can be wiped out in a single day. It was a catastrophic and heartless abuse of the working people of America.

People went hungry. Many small farmers lost their family land. Family stores and family banks went belly up. Proud American workers begged in the streets and lined up by the hundreds to get a bowl of soup. The repercussions, like tsunami waves, spread around the entire planet bringing chaos and suffering to untold millions whose lives were uprooted and permanently changed. Many chose suicide.

By the end of 1930 business failures had reached a record 26,355. Gross national product had stumped 12.6% from its 1929 level. In durable-good-industries especially production was down sharply; as much as 38% in some steel mills, and about the same throughout the key industry of automobile manufacturing, with its huge employment rolls. Despite public assurances, private business was in fact decreasing expenditures for construction; indeed, in the face of softening demand it had already cut back construction in 1929 from its 1928 peak, and it cut still further in 1930. The exact number of laid-off workers remained conjectural; later studies estimated that some four million laborers were unemployed in 1930.[18]

It is in the midst of this cataclysmic depression that Franklin Delano Roosevelt was elected as the third Progressive Party President of the United States. Between 1920 and 1921, America had gone through a brief recession. The government shrank, and in doing so, the economy started up again. But in the depression that began in 1929, the government response was decidedly different with a Progressive Party president in power.

In the brief but devastating recession of 1920–21, government shrank, and the economy healed on its own. Now, with another crisis at hand, FDR had an opportunity to translate Wilson's theory into practice. Roosevelt declared during his 1932 Commonwealth Address that America had hit the end of economic progress and that the job of government now was to "equitably" redistribute its fruits. "The day of enlightened administration [of resources] has come," Roosevelt said.

FDR used his first hundred days in office to seize power from 'tyrannical' business in the name of getting the economy working, largely by executive order. The result was an alphabet soup of agencies designed to regulate and intervene in every aspect of American life.

Under the National Recovery Administration (NRA), Roosevelt created cartels controlled by big business in almost every industry, overseen by the federal government. These cooperatives fixed wages and controlled prices, production, quantities, qualities, and distribution methods under seven hundred competition-killing industrial codes. In its first year alone, the NRA released 2,998 administrative orders approving or modifying existing codes, along with 6,000 press releases, some of which served as legislation.[19]

David M. Kennedy further describes in *The American People in the Great Depression,*

> Almost overnight, NRA mushroomed into a bureaucratic colossus. Its staff of some forty-five hundred oversaw more than seven hundred codes, many of which overlapped, sometimes inconsistently. Cork-makers, for example, faced an array of no fewer than thirty-four codes. Hardware stores operated under nineteen different codes, each with its own elaborate catalogue of regulations. In just two years NRA regulators drafted some thirteen thousand pages of codes and issued eleven thousand interpretive rulings. No matter how constricted their formal legal power, nor how cleverly they strove to exercise what power they had, the mere appearance on the field of that unprecedented bureaucratic horde struck terror into the breasts of many businessmen. "The excessive centralization and the dictatorial spirit," wrote the journalist Walter Lippmann, "are producing a revulsion of feeling against bureaucratic control of American economic life."[20]

The draconian problem with the authoritarianism of this giant federal bureaucracy became evident to the American people, and Congress began to accuse Roosevelt of actually promoting monopolies. This is always the clever ruse of progressivism that in the name of helping the working person, they actually cement the monopolies. There was good reason for this because it was the industrial giants that sat in control of the various code authorities. And subsequently it was small businesspeople who were being adversely affected by the mountain of overregulations. That continues to be the progressive agenda, evident as Obama effectively waged war on the coal industry through such overbearing regulations enacted by the power of his pen through executive orders.

Astonishingly, the New Deal in 1933 seemed to be exacerbating, not redressing, the problem of "balance" in the American economy. "We have been patient and longsuffering," said a farm leader in October 1933. "We were promised a New Deal. . . . Instead we have the same old stacked deck."[21]

Eventually, the legislation that created the NRA was struck down by the U.S. Supreme Court as a violation of the Commerce Clause of the Constitution and also on the grounds that it defied the constitutionally mandated separation of powers that grants the sole right to redact laws to legislators and not to the executive branch through fiat executive orders.

This tactic has not disappeared from the playbook of progressives. It is a problem we faced with Obama when we no longer had a U.S. Supreme Court that believed in the letter of the law in the Constitution. The progressivist concept of the Constitution being a "living charter" that must change with the times provided them with the relativist ability to ignore whatever constitutional law is an obstacle to their centralization scheme.

FDR was outraged that the Supreme Court nullified his Wilsonian attempt to control every aspect of our economic system and hatched a plan to circumvent the court.

In February 1937, fresh off reelection, FDR proposed what would become known as the "court-packing" scheme. Alleging that the Court was flouting the will of the executive and legislative branches in its "activism," he chided the Court for "making law from the bench." He argued that modern conditions demanded action.

His plan called for federal judges to be given the option to resign and accept pension upon reaching the age of seventy. If they refused, FDR would get to appoint additional judges. Since six judges on the Supreme Court were older then seventy, Roosevelt would have the power to potentially expand the Court to fifteen judges, thereby ensuring a favorable majority.[22]

I find FDR's accusation quite distinctly a prime example of Orwellian doublespeak. It is the progressive justices who make law from the bench by ignoring the constitutional law through their relativistic spectacles. It was FDR who transgressed the separation of powers and the Constitution by making law through executive orders. Thankfully, FDR failed to stack the Court, but in the end, he got his wish. Unlike any other president in American history, FDR ignored the two-term rule set by George Washington and was elected to four terms. That enabled him to appoint a grand total of eight Supreme Court justices, which effectively gave progressives a liberal court for decades to come.

The progressive ideology that sought to cure a depression by raising the federal government deficit was a complete and utter failure. It was a total failure then, and it still is today when modern progressives peddle the same lie. Instead of attempting to create fiscal balance, they deliberately increase deficit spending. FDR tripled the taxes in America, and by 1937–1938, the economy collapsed and sent us spiraling into a double-digit depression. It seems that the progressive idea of equality is making everyone poor except the elite and wealthy business cronies who are stuck like giant ticks to the teat of the morbidly obese federal government.

Industrial production plummeted 33 percent. Unemployment increased 5 percent. To add insult to injury, FDR took us off the gold standard and outlawed the private ownership of gold by the American public. The progressive promise of economic recovery through deficit spending and big government sent us deeper into the dark abyss of hunger and suffering. It did not work then. It will not work now. It will never work.

The progressive panacea is but an empty hologram created by slick lies and clever smokescreens. It is the common people who suffered the most. But FDR got his wish—the federal government grew to an enormous power and size. The Wilsonian plan was moved much farther down the track toward the final goal of socialism.

Were it not for World War II, that depression could have permanently wiped out the American middle class. The prosperity that followed the activation of the industrial military complex brought back the middle class by the 1950s to a level of prosperity never experienced before. Unfortunately, that prosperity lulled us into slumber.

But it seems like America simply does not understand the grave danger that lurks in Washington, DC. Once again, in 2008, the voracious greed of Wall Street led to another deliberately engineered debacle. The taxpayers, without their consent, were forced to bail out the rich New York financiers. That caused the Great Recession of 2008, whose artificial name is nothing more than a propaganda attempt to disguise the true nature of the deep depression it created.

Most Americans have no inkling of the true sum of the bailout given to Wall Street. They are only aware of the touted $700 billion that was covered by the media. I have chosen to quote from an article by Mike Collins on the *Forbes* website because it explains this debacle that most of us are completely unaware of. Speaking of the 2008 bailout, he writes:

> Most people think that the big bank bailout was the $700 billion that the treasury department used to save the banks during the financial crash in September of 2008. But this is a long way from the truth because the bailout is still ongoing. **The Special Inspector General for TARP summary of the bailout says that the total commitment of government is $16.8 trillion dollars with the $4.6 trillion already paid out. Yes, it was trillions not billions**

and the banks are now larger and still too big to fail.
But it isn't just the government bailout money that tells
the story of the bailout. This is a story about lies, cheating,
and a multi-faceted corruption which was often criminal.
(emphasis added)[23]

Collins points out that rating agencies such as Standard and
Poor's are actually paid by the banks, which is a direct conflict of
interests. By giving AAA ratings to toxic mortgages, they became
responsible for the catastrophe that unfolded in 2008. I find it quite
ironic that Wall Street, which claims to be the citadel of a free market
system, fell on their knees as socialists demanding a government
bailout. It is as if their underlying motive may be to hold America
ransom and shove socialist policies down our throats.

The operating principles of the big banks are a cesspool
of greed, ethics and criminal intent and they give a very
bad name to free market capitalism. During the housing
bubble Wall street [sic] was considered the heart and soul
of free market capitalism, but when they were in danger of
total collapse they fell on their knees as socialists, begging
the government and tax payers to bail them out.

Many people have asked why the government bailed
them out. Isn't capitalism designed to get rid of the weak
and the failed; so why didn't we just let them fail? The
answer was that they were too big to fail and allowing
them to fail could have created a worldwide depression. In
fact, in a meeting with Congress on September 18th, 2008.
Treasury Secretary Paulson told the members that $5.5
trillion in wealth could disappear by 2pm of that day. In
a meeting with Senator Sherrod Brown, Secretary Paulson
and Federal Reserve Chairman Ben Bernanke said, "we
need $700 billion and we need it in 3 days."[24]

Not only did the taxpayers become the unwitting victims of Wall Street's scam by having to bail them out, but the dastardly betrayal is even more wicked when the bailouts were given with no accountability for how the money was to be used. The greedy merchants gave each other enormous bonuses and laughed all the way to the bank. Meanwhile, the middle and lower class suffered greatly from the debacle as many small businesses went bankrupt, pensions were decimated, and people lost their jobs. In an article in *Rolling Stone* magazine entitled "Secrets and Lies of the Bailout," Matt Taibbi says:

> It was all a lie – one of the biggest and most elaborate falsehoods ever sold to the American people. We were told that the taxpayer was stepping in – only temporarily, mind you – to prop up the economy and save the world from financial catastrophe. What we actually ended up doing was the exact opposite: committing American taxpayers to permanent, blind support of an ungovernable, unregulatable, hyper concentrated new financial system that exacerbates the greed and inequality that caused the crash, and forces Wall Street banks like Goldman Sachs and Citigroup to increase risk rather than reduce it.[25]

Not only did the Federal government lie to us about the true extent of our bailout, but they have done nothing to keep the same thing from happening again. Why should these companies not gamble with high risk speculative financial schemes when they have the taxpayers to pay the price should their risk fail? The very principles of free market competition have been removed, and the Wall Street giants have us as their golden parachutes. The kicker is that none of this information would have been known to us had Congress not forced a one-time audit of the Federal Reserve in November 2011.

The audit not only discovered that the bailout was for trillions, not billions, but it also discovered that the bailout was given to the banks without any requirements or guidelines on how to use the money. Literally, the banks could use the money for any purpose they wished, and not surprisingly they gave each other tremendous pay raises at our expense. In other words, contrary to the free market principles, we rewarded bad management choices with our hard-earned money. It is outrageous, to say it mildly, and criminal, to say it truthfully.

Once again, the merchants used a manufactured economic crisis to consolidate their economic control of our economy. In the wake of the 2008 crash, our banking system became even more monopolized, and the taxpayers helped them do it. The end result was that the big got even bigger at our expense. The government allowed the bailout money to be used by the banks to merge Chase with Bear Stearns, Wells Fargo with Wachovia, and Bank of America with Merril Lynch, further monopolizing the banking industry.

As a result, the twelve largest banks in America now control 70 percent of all bank assets. What is more alarming is that the derivatives that caused the 2008 crash to begin with are once again backed by the FDIC. In effect, we have once again set the banks up to gamble with our financial future. Meanwhile, the monopolies continue to consolidate financial power over our nation. The politicians should be tarred and feathered for their complicity in this crime against "we the people."

The Progressivist New Deal did nothing to stop the monopolies, but instead helped them secure their foothold even deeper. Punishing the guilty and outlawing derivatives is the correct solution. It is Wall Street that backed Hillary Clinton. Meanwhile, our national deficit continues to rise as we bail out the bankers for their chicanery. Americans need to wake up.

The bone that sticks in my throat is that Congress, which is given the responsibility by the U.S. Constitution to be in charge of our financial system, has only audited the Federal Reserve once in 2011.

We have been hoodwinked! What do you mean Congress forced a one-time audit to the Federal Reserve? Congress should have an audit of the Federal Reserve done yearly. What do you mean there are no requirements to these bailouts? This is absolutely outrageous. Every penny given to them in these bailouts should be returned to the taxpayers, not to the federal government to spend on some other trumped up program to grow the federal monster. It should be in a check to all American citizens in proportion to the taxes they paid.

All of this began during Woodrow Wilson's presidency when Congress, under the leadership of Aldrich and against the express writ of the Constitution, signed over its financial power to the private banking institution called the Federal Reserve. Some say that Woodrow Wilson later came to regret having signed this into law and claimed he was deceived into thinking that it would stabilize our economy. Only God knows if he was hoodwinked or a puppet of the robber barons.

What we do know is that the progressive movement has been commandeered by the robber barons, and we the people are being scammed. The Federal Reserve has historically done the exact opposite of what it promised and has allowed the wolf to guard the henhouse. Moreover, today we find ourselves completely at their mercy, for in one fell swoop, they can create another depression like they did in 1907, 1929, and 2008 and reduce our middle class to paupers. America is under extortion.

The New Progressive Agenda

Much of the drive to bust these giant monopolies has now become stymied by the real power behind our elected officials. Whoever controls the purse strings controls the politicians. Is it any wonder that Alan Greenspan, the former chair of the Board of Governors of the Federal Reserve, in his essay "Antitrust" condemns the Sherman Antitrust Act?

> No one will ever know what new products, processes, machines, and cost saving mergers failed to come into existence, killed by the Sherman Act before they were born. No one can ever compute the price that all of us have paid for that

Act which, by inducing less effective use of capital, has kept our standard of living lower than would otherwise have been possible.[26]

The audacity of Greenspan to make such a statement speaks volumes about the mindset of these international bankers and their low view of the intelligence of the common person. While it is true that these large companies may have the potential to create lower prices for goods in the short term, it is nothing more than a smokescreen. They provide lower prices at first until their monopoly is complete. Then historically, they have always raised the prices and gouged the common person.

Even if they would altruistically maintain lower prices (not something I would bet on), the effect of this line of reasoning is anti-American to the core since it is based on rationalizing the unjust measures used to drive honest competitors from their hard-earned businesses. It is no longer free enterprise when only a few are free and the rest are oppressed and robbed.

Such thinking is as wrong as the collectivist ideology, which stipulates that what is good for the collective can be pragmatically rationalized, even if it means an injustice to the individuals robbed of their hard-earned honest money. Our individual rights cannot be sacrificed for the extension of the rights of others. That is the just and proper form of capitalism and the Judeo-Christian worldview.

The overarching concern of the Judeo-Christian worldview is the promotion of true and impartial justice toward all members of society. But the establishment of giant monopolies unjustly robs the right of others to compete in a free market. Hence, it is immoral and illegitimate. If we ever hope to maintain a middle class in America, that must be addressed.

Thus, a just capitalist system should ensure the freedom of opportunity for all investors to compete in a given market and allow for the law of supply and demand to govern prices. The establishment of unfettered giant monopolies short-circuits this lofty goal and impinges on the intrinsic, God-given rights of the smaller investor.

The federal government has no right to pick winners and losers among economic competitors. It has no right to bail out private companies with our taxes. It has no right to endanger our economy by not maintaining a balanced budget.

Because the big banks are not held liable for their crimes, there is no deterrent for immoral transactions such as the sale of derivatives to continue in the future. The cutting up of loans and reshuffling them in derivatives is still legal. That must not be tolerated or we will experience the same catastrophic economic results in the future.

The new progressive agenda is now cleverly masked in the deceitful claim of defending the poor, but it is, in fact, a plan to remove any economic power from the middle class through the socialist agenda.

In May 1966, two Columbia sociologists named Richard Cloward and Frances Fox Piven took to the pages of the iconic leftist magazine *The Nation* to pen an important essay entitled, "The Weight of the Poor: A Strategy to End Poverty."

The idea was astoundingly simple—and sinister: overload the public welfare system at the state and local levels to precipitate a debt crisis that would plunge America even further into poverty. Washington, D.C., would then have no choice but to act and implement a federally guaranteed minimum income level to every American.

"The ultimate objective of this strategy—to wipe out poverty by establishing a guaranteed annual income—will be questioned by some," they wrote. "Because the ideal of individual social and economic mobility has deep roots, even activists seem reluctant to call for national programs to eliminate poverty by the outright *redistribution of income.*"

The Cloward-Piven strategy would overcome those pesky American ideas of individual and economic mobility through a sudden, cataclysmic economic collapse in which millions of people would be forced to become wards of the state, dependent on government for food stamps and basic income. A massive economic crisis would necessitate radical change.

Piven pitched a voter-registration strategy to "radicalize the Democratic Party and polarize the country along class lines," which would be accomplished through collaboration with community-organizing ally the Association of Community Organizations for Reform Now (ACORN), Project Vote, and others over the next two decades.

Cloward-Piven was a resounding success. Beginning with LBJ's "War on Poverty," the number of welfare recipients grew from 4.3 million to 10.8 million in the nine years from 1965 to 1974. Another 100 million Americans now collect some form of check from the government, totaling $1 trillion dollars annually.

The Cloward-Piven strategy succeeded with the urban poor beyond anyone's wildest dreams. It was accelerated with Obamacare and the addition of tens of millions of Americans to federally subsidized health-care programs that American's cannot afford. The next front very likely involves the country's southern border.

We wonder why progressive politicians have no objections to open borders and millions of illegals flooding the welfare rolls and driving up our debt. But we shouldn't wonder because the answer was given to us by Cloward-Piven long ago. After all, an unsustainable federal government is only a bad thing if you believe in the current system.[27]

The progressive plan is rather simple. Overload our national debt by growing the federal government and increasing the welfare rolls through whatever means possible (including massive illegal immigration) until the system implodes. The day the dollar crashes, the progressive social engineers, now controlled by the robber barons, will change our capitalist system of government to a collectivist centralized authoritarian regime.

Checkmate! The end of the middle class and the America Revolution. That will be the day the American dream turns into the American nightmare. That will be the day the New World Order will take center stage.

Notes

1. Sherman Antitrust Act, 15 U.S.C., 1–2, *Cornell Law School, Legal Information Institute*, https://www.law.cornell.edu/uscode/text/15/chapter-1.

2. "Antitrust," *Cato Institute*, https://object.cato.org/sites/cato.org/files/serials/files/cato-handbook-policymakers/1997/9/105-39.pdf.

3. *CATO Handbook for Congress*, https://object.cato.org/sites/cato.org/files/serials/files/cato-handbook-policymakers/2003/9/hb108-38.pdf.

4. Ibid.

5. Glenn Beck, *Liars: How Progressives Exploit Our Fears for Power and Control* (New York: Threshold Editions/Mercury Radio Arts, 2016), 27–28.

6. Ibid., 31–33.

7. Ibid., 48–49.

8. U.S. Constitution, Article 1, Section 8, *Cornell Law School, Legal Information Institute*, https://www.law.cornell.edu/constitution/articlei.

9. Simon Johnson and James Kwak, *13 Bankers: The Wall Street Takeover and the Next Financial Meltdown* (New York: Vintage Books, 2011), 20.

10. Ibid., 21.
11. Ibid., 22–23.
12. Ibid., 23.
13. William McKinley, "First Inaugural Address of William McKinley," March 4, 1897, *Yale Law School Lillian Goldman Lawa Library*, http://avalon.law.yale.edu/19th_century/mckin1.asp.
14. "The Assassination of President William McKinley, 1901," *EyeWitness to History*, http://www.eyewitnesstohistory.com/mckinley.htm
15. Glenn Beck, *Liars: How Progressives Exploit Our Fears*, 54–55.
16. Simon Johnson and James Kwak, *13 Bankers: The Wall Street Takeover and the Next Financial Meltdown*, 27–28.
17. Ibid., 30.
18. David M. Kennedy, *The American People in the Great Depression: Freedom from Fear:* Part 1 (Oxford, UK: Oxford University Press), 58–59.
19. Glenn Beck, *Liars: How Progressives Exploit Our Fears*, 91–92.
20. David M. Kennedy, *Freedom from Fear: Part 1*, 185–186.
21. Ibid., 190–191.
22. Glenn Beck, *Liars: How Progressives Exploit Our Fears*, 96.
23. Mike Collins, "The Big Bank Bailout," July 14, 2015, *Forbes*, https://www.forbes.com/sites/mikecollins/2015/07/14/the-big-bank-bailout/#1a2662ee2d83.
24. Ibid.
25. Matt Taibbi, "Secrets and Lies of the Bailout," January 4, 2013, *Rolling Stone*, https://www.rollingstone.com/politics/politics-news/secrets-and-lies-of-the-bailout-113270/.
26. Alan Greenspan, "Antitrust," (New York: Nathaniel Branden Institute, 1962), http://keever.us/greenspanantitrust.html.
27. Glenn Beck, *Liars: How Progressives Exploit Our Fears*, 155–156.

CHAPTER 8

● ● ●

A JUST AND BALANCED
CAPITALIST SYSTEM

My friends, the War of Independence is over, but not the American Revolution. If we are to preserve this great nation of ours, we must educate our children in order to inoculate them from the smokescreens being used to indoctrinate them. We must understand that freedom is never free. We must fight to preserve it for every generation or the Enemy of Man will take it from us. As Ronald Reagan said, we are always only one generation away from tyranny.

We must elect politicians who are constitutional conservatives, and we must oppose the collectivist-globalist agenda at every level, from the local government to the federal government. We must involve ourselves in the political system or pay the consequence of our apathy. We must strive to restore a just and balanced form of capitalism or pay the consequence of becoming a slave of the state. There is no other choice if we are to pass on to the next generation the freedoms we have been privileged to enjoy.

Justice Demands a Share in the Profits between Labor and Management

There is yet another consideration that must be protected in order for a healthy and balanced form of capitalism to remain efficient and fair. Justice is not met if the welfare of the employee and the consumer is not considered.

Giant strides were made in our country in the last century to secure the rights of the employed to unite in bargaining unions to ensure a more equitable share in the profits of the industries for which they work and for a safer work environment. We have already mentioned the greed of the industrialist elites to maximize their profits at the expense of those they exploited. And in the process, they created inhumane work conditions with dismal pay.

If you are familiar with our nation's history in the past century, you know that children were sometimes chained to machines, working for 12 hours per day. These machines would often cause severe injuries to workers. Many industrialists were not concerned with the safety of their workers but only with maximizing their own profits. Attempts by workers to unite in order to demand justice were met with violence and often lethal consequences by hired thugs.

That was the fuel that fired the collectivist agenda throughout the world from progressivists to socialists to communists. But I ask you to consider the historical evidence: Those who were hoodwinked by the collectivist ideology have not brought greater freedom and justice to the common person. Quite the contrary; they have established ruthless authoritarian central governments that have left oceans of innocent blood in their wake. But I need not elaborate on these injustices since they are well known and documented by others. Greed has ever been the culprit that brings us to heartless abuse and violence. But it cuts both ways.

If workers begin to demand a higher wage than what can be reasonably absorbed by a company, they will run the company into bankruptcy. A labor movement that disregards this fundamental logic and continues to press for more concessions than the market can absorb is biting off its nose to spite its face. Those unions, armed with collectivist ideologues, have that goal in mind to destroy capitalism.

They may end up getting what they think they want and losing what they really need. Workers may get a temporary raise but then lose their jobs as the company goes under. The exploitation of greed has ever been the work of devils eager to grant Faustian contracts. They work on both sides of the fence.

The enemies of capitalism have targeted this potential weakness in a free market economy from the beginning. It is fair to say that practically all the labor unions have been targeted and infiltrated by the communists as the ideal weapon to strike against capitalism from the inside. Their goal has always been to cause an economic collapse in order to strike the match of revolution.

The incitement of class warfare in a nation can bring its economic output to a trickle and lead to armed rebellion. This has been the modus operandi of communist revolutions around the world. Do not for one second think that America is above that.

It is precisely because Lenin was able to bring attention to the giant monopolies that were sprouting all over the industrialized world that he was able to garner support for communist revolutions everywhere. His message that capitalism had died was the fuel that fired the proletariat to rise up with arms, promising the creation of the dictatorship of the proletariat.

> Free competition is the fundamental attribute of capitalism and of commodity production generally. Monopoly is exactly the opposite of free competition; but we have seen the latter being transformed into monopoly before our very eyes, creating large-scale industry and eliminating small industry, replacing large-scale industry by still larger-scale industry, finally leading to such a concentration of production and capital that monopoly has been and is the

result: cartels, syndicates and trusts, and merging with them, the capital of a dozen or so banks manipulating thousands of millions. At the same time monopoly, which has grown out of free competition, does not abolish the latter, but exists alongside it and hovers over it, as it were, and, as a result, gives rise to a number of very acute antagonisms, frictions and conflicts. Monopoly is the transition of capitalism to a higher system.

If it were necessary to give the briefest possible definition of imperialism we should have to say that imperialism is the monopoly stage of capitalism.[1]

It must be noted that it has been the excesses of these giant industrial companies that have historically created the backlash of the proletariat that for so long has bled our planet through countless violent revolutions. But neither extreme creates a just government for the common person. Communism, socialism, and supercapitalism are equally hardwired to exploit the proletariat. All of them will lead to tyranny.

Lenin's Soviet Union is now dead because of the failed economic policies of collectivism. Today, Russia is but another oligarchy led by the old KGB of the Soviet empire. It is a ruthless mafia that has little regard for the welfare of the Russian worker. But its colonialist appetite has not been muted. President Vladimir Putin has set his eyes on regaining much of the territory the country lost. The panacea and utopia imagined in Lenin's mind was nothing more than a hologram created by the inability to understand the true nature of humans according to the Judeo-Christian worldview. Centralized power is the indisputable perfect recipe for the abuse of the common person. That is a historical fact, not theory.

There is an alternative to collectivism and the monopoly of supercapitalism. From a Judeo-Christian perspective, I envision the use of the tools of capitalism in such forms as offering real profit-sharing incentives to workers. That would ensure a more equitable distribution of the fruits of labor and of the capital that has been produced to enrich both the laborer and the employer. In other words, it would create a more just compensation for labor, as well as for the capital risk ventured. Moreover, it would maximize the quality and quantity of the product since laborers would have a vested interest in maximizing their own profits.

But for this profit sharing to be workable as an incentive, it must be enough to really be an incentive and not just some corporate gesture to be used for propaganda. My wife is a flight attendant for a major airline. She recently received a profit-sharing check for $.20. The cost of the postal stamp was more than the amount of the check. Yet that company registered record earnings that year. A small profit sharing check is nothing more than an insult to the company's employees.

Justice demands that the individual risking the most should be recompensed the most. But the balance between the recompense of the individual venturing capital and the laborer, who creates the return on the capital, should be fair. If the laborer to some extent shares the risk of the loss of profit, then to that extent, he or she should be rewarded when there are profits. Some may argue that the laborer does not share the risk of the loss of profit. But that is not true. Laborers may have their wages reduced, their benefits curtailed, or in the worst-case scenario be laid off or lose their jobs altogether. Their risk is real.

What is more, the competition between companies in any industry stimulates the efficiency of production and the quality of the product itself in order to capture a larger slice of the market, and so it serves to improve the economy as a whole. But that slice should not be allowed to become a monopoly since monopolies short-circuit all of it.

Supercapitalism thus serves only to increase the gulf between those who have and those who have not, which inevitably enslaves the common person. But socialism and communism do not give us a better alternative. Their political and economic philosophies are skewed and unjust, which result in the eventual abuse of the proletariat.

Millennials who have been angry at the monopoly of Wall Street are unknowingly choosing to empower them even more by supporting progressivism and socialism. It is their fastest route to fleece the flock to the bare skin and end the middle class altogether.

I was told by one of my college professors that I was not a realist when I shared with him my view regarding our constitutional form of government and the Judeo-Christian perspective of a balanced form of capitalism. He, of course, was a humanist and socialist who believed that only by establishing a strong central government and forcing the redistribution of wealth could justice be accomplished for the common person. He believed that humanity was basically good and that only through a strong government could humanity evolve into a more just society. I explained my thinking:

> It is you who is not a realist. Your blind faith in the goodness of man has been shown to be nothing more than a hallucination by every single country that has adopted the centralization of power and turned to collectivism. The loss of individual freedoms and the natural oppression that comes from centralizing power is the hallmark of the bankrupt worldview that justice could ever be attained by a system of relativism that leads only to pragmatism and the abuse of the powerless.

Needless to say, I was not his favorite student.

The Disintegration of the Law of Supply and Demand

A relativist-Darwinian-collectivist political ideology cannot produce justice and fairness in a society since it conversely promotes the survival of the fittest and legitimizes the strong preying on the weak. But neither has the supercapitalist political ideology produced justice and fairness since it also promotes the Darwinian axiom of the survival of the fittest and refuses any moral constraints on an unfettered greed for power. Both are driven by a globalist-imperialist drive for total power over the entire world.

The balance of the interplay between supply and demand, which is the bedrock of a true and balanced capitalist economy, has as its basic component to be efficacious, the necessary competition between fabricators or suppliers to produce goods for consumers at a competitive price. If the very wealthy are allowed to set up monopolies, cartels, or syndicates by which they are able to control all the suppliers of a given product, then they are able to unjustly raise the price of goods to an unreasonable level. We have seen this with the price of oil. We are at their mercy. They can raise the price as often as they want, and we have no choice but to pay it if we want to drive to work and buy groceries.

Or if the very wealthy are allowed to purchase raw materials for a given product at a much-diminished bulk rate, which allows them to undercut all other competitors, then it is only a matter of time before a monopoly is established. That would inevitably drive the masses, which are dependent on those products, to an ever-lower financial stratum as they are systematically gouged of their hard-earned money. Each year, common people are paying more and more for monopolized items and have less and less to save or invest for the future of their progeny.

We have seen this in the soaring rise of the price of gasoline. Gas prices are elevated artificially and the consumer has no recourse but to buy it at whatever price it is sold. What choice do we have if we are to get to work and drive our children to school?

The oil industry functions as a single cartel, exhibiting only a few cents difference between the few companies that share this energy monopoly. There is no real competition among them. Prices are artificially raised at whatever level they think the public will swallow without revolting and in response to the rise in the cost of a barrel of crude oil. They do not lose any money when the price per barrel rises. On the contrary, the scarcer the oil is, the more they can charge and the more profit they make.

But this process is not reserved just for the oil industry. It applies across the board, even in retail and manufacturing goods industries. It begins first by establishing a large enough company that they are able to buy raw materials in cheaper bulk rates. That allows large companies to create the product at a much cheaper price. They then lower their prices to the absolute rock bottom level, which undercuts small businesses, which are unable to compete or make a profit.

In addition, they may also saturate the economy with the overproduction of that commodity, thus creating a glut that drives prices down further. The giant companies can absorb the temporary loss in revenues until the small businesses are obliterated. Inevitably through this process, they eventually establish a virtual monopoly.

The small investor invariably goes out of business, incapable of competing. And when giant companies establish a sufficient monopoly over that specific market, they raise their prices without fear of any competition.

It is, therefore, my opinion that justice demands that our government should not take on an absolute laissez-faire position in this matter. The job of the government is to protect the rights of individuals and corporations. But we must note with great gravity that the burden of needless overregulation is another method that monopolies use to exterminate competition. Such overregulation will stifle the growth of small businesses and interfere with their rights.

In the last few years, we have seen socialist-progressive elements in our federal government that have calcified the arteries of our small and medium-sized businesses with overburdening mounds of needless regulations that have sent those with the financial ability to pack up and go overseas. Millions of jobs have been lost to Mexico, India, and China thanks to these strangling regulations.

No individual has the right to interfere with the rights of other individuals just as no corporation has the right to interfere with the natural rights of other corporations. And the government has no right to interfere with the rights of corporations. Understand this: Free enterprise does not mean anarchical enterprise. In order for freedom to exist, the boundaries of justice must be enforced. The job of the government is simply to ensure justice, and all regulations should be confined to this goal.

The price of raw materials should be set according to the cost of production and a fair profit value. The free market economy will provide the checks and balances necessary to set its price, but the government should require that this price should be uniform for all corporations competing. The government should never step in to interfere in this regard unless an injustice warrants it. But this system cannot work without competition, for if the supplier is a monopoly, it can raise prices unfairly.

Our country has been founded on the principle of equal opportunity for all. It would be considered unjust if the price of a shirt were $10 for a white person but $20 for a black person, or if the price of any item were lower for one sector of our society and higher for another. Justice demands that the price of materials should not vary in bulk sales, giving the small investor equal opportunity to succeed. Not doing so will inevitably lead to a monopoly. It is inevitable.

Congress should reinstitute strict antitrust laws that were neutralized during the Great Depression right before the Second World War. Failure to act upon this will ensure that our nation's economy will be completely controlled by a few elite industrialists and financiers. In the end, consumers will pay the price if these measures are not heeded, and elite financiers will create a shadow government behind our government through their economic power.

We have already seen this happen in the food industry as family farms have been systematically run out of business and giant food manufacturing conglomerate industries have slowly but surely taken over nearly every aspect of the production, distribution, and sale of food in our country.

The control of our food supply in the hands of a few is a mighty dangerous weapon to wield! We should provide an equal plane of competition for the small farmer who could provide for us the localized food source that eliminates enormous transportation costs. Today, food travels an average of 1,500 miles to reach consumers. The enormous cost for fuel to transport food adds to the individual's cost of living and to the national debt since we are forced to purchase this fuel from countries that are less than friendly to us.

The localization of food production must take center stage in order to reduce our negative trade deficit to oil-rich countries. Moreover, these farmers can help grow renewable fuel sources that we desperately need to establish an independent economy that is free of the Islamic oil cartels.

At the moment, the gulf between the very wealthy and the masses continues to increase, not because the masses are not willing to work and produce but because they are forced to pay inflated costs for these monopolized items and are unjustly kept from being able to compete in the free market.

This invariably causes them to spiral downward in their living standards as the buying power or value of their money is diminished. The answer to the farmer's problem is not government subsidy that increases the tax burden and increases our national debt. The answer is control of monopolies and the establishment of just and equitable purchasing rights.

Our federal government should be reduced to the smallest size possible in order to effectively comply with its constitutional mandate. Power should be decentralized wherever possible. One of the primary brakes that has ground our economy to a halt has been the overburdening cost of the overinflated federal government that has transgressed its constitutional boundaries and is attempting to circumvent the checks and balances enumerated in our Constitution.

We should demand that our government have a balanced budget by passing a constitutional amendment that secures this essential element for the future existence of the great middle class and our sovereignty as a nation. Illegal immigration laws should be enforced to reduce the welfare explosion that will bankrupt our nation. If we allow our national debt to continue at this rate, the dollar will collapse, and with it will go our Constitution and the Bill of Rights.

We should demand that we return to the gold standard or at least a bimetallic standard. If we do not, our dollars will one day not be worth the paper they are printed on, our retirement savings will go up in a puff of smoke, and the middle class will be a thing of the past.

We should push for the right of the president of the United States to have line item veto power to strike the pork belly parasites that plague every good piece of legislation, which subsequently causes our national debt to continue to increase through an ever-burgeoning bureaucracy. The president of the United States should make these parasites public who attempt to ride on the backs of necessary legislation so the public may cry out against these private lobbyists.

Sadly, we are crossing the line from capitalism to supercapitalism in our country with ever-increasing momentum. If we do not oppose it, we will see the death of the middle class in America. Soon it will be too late.

So, then, we see that supercapitalism is like a cancer cell that begins with a healthy host and grows in an unnatural and grotesque way, killing the very host organism that gave it life. In my mind, it is every bit as evil as collectivism and communism, for it divests all the wealth of a nation not on its laborers but on the elite living extravagantly off the sweat and toil of exploited workers. Supercapitalism is no different from the tyrannical monarchies that enslaved our forefathers for so long and that our nation rejected and fought a revolution against to break those shackles.

We are witnessing the beginning of a new nobility, a new aristocracy, and if we do not take the steps necessary to stop it, our nation will soon find itself full circle under despotic control by another tyrannical King George, but he will be a merchant, perhaps of the digital age.

The very same condition that existed in the monarchial aristocracies of old that led to the inevitable violent rebellions of the downtrodden masses will, no doubt, if left unchecked, surely lead to bloodshed in the future if supercapitalism is also left unchecked. But it need not come to this. Today, we still have a brief window of opportunity to act. If we do not, we have no one else to blame.

The Problem of the Human Condition

The principles that brought about the concept of a government that represented fairly all individuals within a society (a Judeo-Christian constitutional republic) and protected them from the powerful through equal access to justice and opportunity were the natural outworking of the principles of the Word of God. No other religious philosophical presupposition in the world could have developed this form of government.

But unfortunately, even this form of government is today being corrupted, and the prosperity that created our enormous middle class, which naturally emerged from these liberties, is in danger of once again being swallowed up by the wealthy elite; that is, the new global aristocracy created by supercapitalism is consuming it, bit by bit.

The sad reality of the human condition is such that the centralization of power will inexorably lead to a corrupt tyranny that callously exploits the masses. This is the direct consequence of the fall of humankind in the Garden of Eden, which has marred our human nature. It is this realistic view of humanity that requires a system of checks and balances.

Within every person is the potential to do great good, but there is also the potential for great evil. As Solzhenitsyn so insightfully noted, the line between evil and good is not drawn between one nation and another. It is not drawn even between one human and another, but within the heart of every soul:

> If only there were evil people somewhere insidiously committing evil deeds, and it were necessary only to separate them from the rest of us and destroy them. **But the line dividing good and evil cuts through the heart of every human being. . . .**
> During the life of any heart this line keeps changing place; sometimes it is squeezed one way by exuberant evil and sometimes it shifts to allow enough space for good to flourish. One and the same human being is, at various ages, under various circumstances, a totally different human being. At times he is close to being a devil, at times to sainthood. But his name doesn't change, and to that name we ascribe the whole lot, good and evil (emphasis added).[2]

That is reality. If we ignore it, we will pay the price for it. There are some cases where that line has been pushed so far and for so long to the side of evil that no good can be seen in a particular person. Those who have given themselves over to the dark side out of selfish interests have effectively suppressed any sign of good in them and have simply become human monsters.

These greedy few manipulate the masses by enforcing the magnificent lie on us and sending us merrily on that elusive chase for the mirage of self-sufficiency and self-healing at the rainbow's end. But the vast majority of us fluctuate back and forth, never rooting for one side or the other completely.

We fail to understand our true identity and the redemption that is offered freely by our merciful God. That alone can make us whole. That alone can heal our soul. That alone can keep that line to the side of selflessness through the power of the Holy Spirit in us.

Of this one thing you can be sure: The power of the dark side lurks just below the surface, waiting for the opportune moment to strike within each one of us. This is the very real reflection of humanity due to the fall of humanity in the Garden of Eden.

We are not all evil, and we are not all good. We all have the capacity to do both good and evil. Nevertheless, without the redemptive power of God's grace, we are powerless to overcome the evil in us through our own self-efforts. Only when we humble ourselves and by faith admit our failings does His grace shower us with supernatural power to overcome the evil in us. We must understand that the enemy without is also the enemy within.

However, we are not worthless "zeroes," as Francis Schaeffer would say. We are infinitely valuable, and our lives do have transcendental significance, even when our merciful creator has not redeemed us through our profession of faith in Him. Therefore, the sacred dignity of a single human life is not compromisable. It rests not on a person's utilitarian value to the state but upon a person's descent from the breath of God when He created that person.

And yet our form has been marred from its original glory and intent. We are not as we were meant to be. We have chosen the wrong path of selfishness. Evil lurks below the surface of the skin of every human being on the face of our planet. No honest person can deny the truth of our fallen nature.

For that reason, our Founding Fathers, imbued with a clear understanding of the true human condition, sought to establish in our government a system of checks and balances. Power was divided and decentralized in order to forestall the inevitable attempt by the greedy to take away our basic freedoms that were endowed to us by our creator.

These rights have been protected, although imperfectly, by our Constitution, for that was the genius of our founders' intent and design. If we myopically disregard that basic truth, it will result in dire consequences for our posterity in the future.

The unparalleled genius of dividing the judicial, executive, and legislative branches of our government has served to keep our nation free of tyranny by not allowing the power of the government to rest in a centralized form. Even the legislative branch was divided into two houses, reflecting the English House of Commons and House of Lords. The House of Representatives reflects the will of the common person, and the number of representatives in each state is determined by population density. The Senate, however, is comprised of only two members for each state, representing the rights of the individual sovereign states.

The Need for Election Reform

Unfortunately, the evolution of the modern election process, through slick television commercials, has led to a near monopoly by the financial elite since their monetary power is the crucial element necessary to fund the incredibly expensive election campaign of any aspiring politician.

Simply put, in order to be elected to office, a person must expend a great deal of money for advertising. And the winner in the overwhelming majority of cases is not necessarily the one who is best suited or qualified but rather the one who has the slickest commercials and is best funded.

Adding to this travesty is the fact that an elite few own the major mass media networks, which slant the news and the political coverage in such a way that it is more propaganda than news. Gone are the days of serious journalists who sought to provide unbiased news to American viewers. The control of the major airways by those who uphold the globalist agenda is the death toll of the democratic process envisioned by our forefathers. Unbiased newscasts are but a faint memory in the history of America. The unbridled progressivist agenda they brashly spout is no longer dissimulated. Newscasts are radically biased and completely opinionated, serving more like Himmler's propaganda machine than a news station:

> The media and especially television have indeed changed the *perception* of not only current events, but also of the political process. We must realize that things can be presented on television so that the *perception* of a thing may be quite different from fact itself. Television not only reports political happenings, it enters actively into the political process. That is, either because of bias or for a good story, television so reports the political process that it influences and becomes a crucial part of the political process itself. A good example was Walter Cronkite's part in orchestrating the Gerald Ford candidacy for Vice-President at the 1980 Republican Convention.
>
> We must realize that the communications media function much like the unelected federal bureaucracy. They are so powerful that they act as if they were the fourth branch of government in the United States.[3]

This remark made more than 30 years ago only scratches the surface of what is going on today. The bold disregard even for the appearance of a balanced view makes our major television networks the private public relations companies of their select politicians who pander their globalist agenda.

Campaign reform has therefore been the topic of every presidential election I can remember. But there is still no substantial action in this regard. And I do not expect it anytime soon unless the public is awakened from its lethal slumber.

Political campaigns should not be funded by private industry. All candidates of a legitimate political party should be given equal time on television and radio, and in newspapers that will equally declare their political platforms.

That should be funded by our tax dollars, and then the public will be motivated to choose a candidate upon his or her platform and not by the effectiveness of slick Madison-Fifth Avenue commercials.

The way our election process is managed today, the public is influenced not by the meat of someone's platform but by the person's ability to debate or a superficial persona. The fact that an individual has superior debating skills does not in any way qualify him or her for leadership. The character and integrity of the individual need to be considered when choosing someone for a position of responsibility. Our decision needs to be made according to the nuts and bolts of a candidate's platform and the qualifications of the individual, not on the skills of their speechwriters or the bias of debate moderators.

The powerful lobbyists in Washington, DC, have for too long sequestered our political representatives and perverted the political system intended by the framers of the Constitution. Only a grassroots movement to override these forces will produce the real change necessary to return our nation to the system it was designed to have.

Unfortunately, there will not be true and just equality in this world until the Messiah reigns from Jerusalem. But the day is coming when justice will prevail and peace will reign supreme as He reigns in a regenerated planet in the true "Age of Aquarius."

Nevertheless, it is our responsibility to resist the forces of darkness and not roll over out of apathy, laziness, or fear, which would give Satan his tyrannical global kingdom unopposed. This is our task in the battle of the long defeat in every generation. To do anything less is to help the Enemy of Man.

Our world is quickly sinking into this great crevasse that Solzhenitsyn warned us about. The paradigm shift is now almost complete, and the matrix of reality for most nations is the evolutionary, Hegelian concept of reality.

The application of the evolutionary matrix to the relationship of nations by its very nature will manifest itself in war and the quest to dominate one another. How can it become anything less? It will be justified by their sacred evolutionary mantra, the survival of the fittest.

What will become of a planet whose people have been raised under the worldview of the evolutionary paradigm? What will become of a generation that has been taught that there is no right or wrong choices? What will come of a people who have been indoctrinated to believe that the matrix of reality is the survival of the fittest? Solzhenitsyn answers this for us with clarity, foresight, and great wisdom: "The truth is that one day they will turn and trample on us all. And as for those who urged them on to this, they will trample on them too."[4]

But most of us, insulated in this great and prosperous country of ours, have great difficulty understanding the sheer gravity of the loss of freedom in nations with such an underlying worldview. We are not so deep yet in the crevasse that light cannot still enter from above. But soon there will be nothing but utter darkness. And I am afraid that perhaps words may not be enough to communicate such a truth.

Perhaps we must first experience the loss of our freedoms before we can truly understand and appreciate their value. Those of us who have lived in other countries sometimes find it frustrating to attempt to communicate that truth to the relatively wealthy, satiated, and complacent nations of the West. I sense that Solzhenitsyn had that same frustration when he tried to warn us of our pervasively entrenched complacency. It is the poison that will kill our liberties.

The West has great difficulty relating to a society that does not have the freedoms we take for granted. It is almost impossible for anyone in our culture to feel the oppression that exists in nations that have moved away from the Judeo-Christian mandate of individual freedom.

In those nations, a simple innocent statement can be turned against you with lethal consequences for you and your entire family.

And in their skewed sense of justice, such a statement can condemn a person to imprisonment or even death. Solzhenitsyn recounts such an incident in the Soviet Union:

> In the first volume of *The Gulag Archipelago* I tell of an event which was recounted not by some insignificant arrested person but by all the members of the Supreme Court of the U.S.S.R. during that brief period when I was in the good graces of the regime under Khrushchev. A Soviet citizen had been in the United States and on his return said that they have wonderful roads there. The KGB arrested him and demanded a term of ten years, but the judge said: "I don't object, but there is not enough evidence. Couldn't you find something else against him?" So the judge was exiled to Sakhalin because he dared to argue, and they gave the other man ten years.[5]

It is hard for an American to understand the true gravity of the injustices that are daily occurring in such oppressive governments where individual rights are nonexistent because of their atheistic, pantheistic, or Islamic form of government. Every day in countries throughout the world, these oppressive monsters murder people merely because they do not think along accepted lines or because they believe in another god.

We have seen this by the millions in the purges of the Soviet Union and China, but now also with Muslims. It is not enough to just believe in any deity. Those who reject the Judeo-Christian worldview in favor of Allah and the message of Muhammad are no less barbaric than atheists. If their deity is a deity of war and hatred such as Allah, the colossal consequences to human dignity and freedom are just as disastrous.

Countless men and women are taken from their homes and then disappear into the night. Jewish children are being killed in Israel, simply because they are Jews. The Muslim call to annihilate Israel and every Jew from the face of the earth is a horrendous indictment against the Muslim religion of hatred and violence.

In countries where Islam is the state religion, the mandates of Sharia law are forced upon all people. Those who teach or believe any other religion are charged with apostasy and subsequently killed. There are no individual human rights, and people's rights are determined by their spiritual dictator. Women are regarded as simply personal property. In some Muslim countries, they are not even allowed to receive an education.

In countries where naturalism is the religion, pragmatism becomes the rule of law, and their skewed sense of justice cannot be challenged. Moreover, what is pragmatically beneficial to the government outweighs the individual rights of citizens. It is hard for us to understand that this is where we in America are headed if we abandon the Judeo-Christian worldview that engendered our great nation.

Gone will be the stabilizing force of the great middle class that was spawned by the application of the Judeo-Christian worldview in our American political and economic system. The wealthy and political elite will hoard all power, and the slave caste will serve only to better the lot of the elite.

The great freedoms afforded by our Judeo-Christian-inspired American republic spawned a middle class as never before experienced in any other nation. But our foundation is being eroded, and the middle class is being systematically reduced as Satan continues to move the chess pieces that bring us to the globalization process and their carefully contrived socialist kingdom of his beloved Antichrist or the autocratic tyranny of an Islamic Mahdi. Which side will prevail? Only time will tell.

The Neo-Nobility

The greatness of a nation is measured principally not by the greatness of its politicians or generals, not even by its economic

power and wealth, but by the moral fiber of its citizens. It is not a backbone of rubber created by relativistic mores but a rock-solid backbone created by the Judeo-Christian absolute standards revealed to humankind through God's propositional revelation. When that moral fiber is strong, the rest comes together.

That is what produces sound, healthy families that can then produce great leaders, just governments, and a thriving economy. But our moral fiber is being sabotaged by the relativism being taught in our schools, movies, songs, television programs, and news as well as magazines and newspapers. Our representative form of government that was designed to provide elected representatives for all classes of our culture has been effectively torpedoed.

The rich elite, through the well-conceived plan and executed tactic of the formation of the Federal Reserve, has established an economic power structure behind our elected governmental officials. Our politicians have spent our hard-earned money beyond the means of the country and will continue to do so unless we stop them.

They have created a gargantuan and skyrocketing national debt that leaves us at the mercy of these international industrialists who own and operate the Federal Reserve. These wealthy international industrialists have literally bought our nation, stripped it right out of our hands, as we have become hopelessly indebted to them.

Our presidents as well as our senators and representatives are no longer elected by the people at large. The wealthy corporate elite essentially buy them before they even run for office. Without the enormous sums of money necessary to sell the candidate through sleek TV advertisements, there is not a snowball's chance in hell at being elected to any state or national governmental seat.

The American ideal that fostered the American Revolution to balance the power of the wealthy nobility so the common person could live in a just form of government that would ensure their financial, civil, and religious liberty has been effectively defeated through subtle subterfuge.

An even more dangerous neo-nobility has formed a Washington shadow cartel that now rules America. They are the swamp creatures who use our intelligence services against our own citizens. They may not carry grandiose titles or dress in scarlet robes and gold crowns, but they are, indeed, a new nobility—a neo-nobility, an extremely powerful and sinister nobility—that cleverly remains in the background while holding the real reins of financial and political power and the destruction of all who stand in their way. Their agenda has been to steadily centralize power in the federal government.

Little by little, they have effectively eroded the financial power of the middle class. Little by little, the essential elements of our economy are being consolidated in vast conglomerates that will hold us ransom when the right moment arrives. Little by little, they are paving the way for the nationalization of our industry.

They started with the seemingly benign takeover of our nation's water supply. Private property owners were stripped of the rights to any natural lakes, rivers, or creeks on their farms or ranches. Little by little, they are moving us toward a socialist government. And all the while, we have been silenced by our myopic obsession for the gratification of our fleshly desires in the present. We have sold our birthrights for a pot of porridge.

Today, only the very wealthy—or the pawns of the very wealthy—can compete for any elected office of any real national power. If we are to preserve our nation, we must push for real election reforms. Our second president, John Adams, in his inaugural address in the city of Philadelphia on March 4, 1797, clearly delineated the very heart of our American Revolution:

> What other form of government, indeed, can so well deserve our esteem and love?
>
> There may be little solidity in an ancient idea that congregations of men into cities and nations are the most

pleasing objects in the sight of superior intelligences, but this is very certain, that to a benevolent human mind there can be no spectacle presented by any nation more pleasing, more noble, majestic, or august, than an assembly like that which has so often been seen in this and the other Chamber of Congress, of a Government in which the executive authority, as well as that of all the branches of the Legislature, are exercised by citizens selected at regular periods by their neighbors to make and execute laws for the general good. Can anything essential, anything more than mere ornament and decoration, be added to this by robes and diamonds? Can authority be more amiable and respectable when it descends from accidents or institutions established in remote antiquity than when it springs fresh from the hearts and judgments of an honest and enlightened people? For it is the people only that are represented. It is their power and majesty that is reflected, and only for their good, in every legitimate government, under whatever form it may appear. . . . If national pride is ever justifiable or excusable it is when it springs, not from power or riches, grandeur or glory, but from conviction of national innocence, information, and benevolence.

In the midst of these pleasing ideas we should be unfaithful to ourselves if we should ever lose sight of the danger to our liberties if anything partial or extraneous should infect the purity of our free, fair, virtuous, and independent elections.[6]

Indeed, there is nothing more sublime to a benevolent human mind than a form of government that rests its authority not in the partial interests of a select wealthy class but in the common people for the welfare of all. His warning was prophetic.

We have lost sight of the very kernel of our American Revolution. We have been unfaithful to ourselves to allow an elite few to pervert and deform the ideals of our Founding Fathers. There are still those with great resolve who have fought and will fight to secure this American dream, but the wind bodes dark and ominous times on the near horizon. And although our War of Independence is over, the American Revolution is not.

But is it possible for Americans to really understand the gravity of their situation? Is it possible to wake a people so deep in slumber before the hammer falls? Listen again to Solzhenitsyn's warning:

> Is it possible or impossible to transmit the experience of those who have suffered to those who have yet to suffer? Can one part of humanity learn from the bitter experience of another or can it not? Is it possible or impossible to warn someone of danger?
>
> How many witnesses have been sent to the West in the last sixty years? How many waves of immigrants? How many millions of persons? . . . Coming from different countries, without consulting with one another, they have brought out exactly the same experience; they tell you exactly the same thing; they warn you of what is now taking place and of what has taken place in the past. But the proud skyscrapers stand on, jut into the sky, and say: It will never happen here. This will never come to us. It is not possible.
>
> It can happen. It is possible. As a Russian proverb says: "When it happens to you, you'll know it's true."
>
> But do we have to wait for the moment when the knife is at our throat? Couldn't it be possible, ahead of time, to assess soberly the worldwide menace that threatens to swallow the whole world? I was swallowed myself. I have been in the dragon's belly; in it's red-hot innards. It was unable to digest me and threw me up. I have come to you as a witness to what it is like there, in the dragon's belly.[7]

The great prophetic question that Solzhenitsyn has posed is quite deep: "Is it possible or impossible to transmit the experience of those who have suffered to those who have yet to suffer?" He did not say "those who have not suffered"; he said "those who have yet to suffer." He knows that if we do not learn from the suffering of others, we will condemn ourselves to certain future suffering.

For the first time in our American history, we have a candidate for president—Bernie Sanders—who is not hiding behind the progressivist veneer but has publicly declared himself a socialist. Even more concerning is that practically every freshman Democrat in Congress is locked in step with Bernie's socialist agenda. The result of our modern public educational system has not only dumbed down our children but also primed them to accept the socialist agenda, lured by the grand illusion of the promised cornucopia, the promise of free stuff. The entitlement mentality has spread like a contagious virus in the minds of our youth. Gone is the Judeo-Christian work ethic that made our country the most powerful economy in the world. If we fail to confront this now, our once great nation will be infected by the lethal parasite of socialism that will destroy us from the inside.

But our danger comes also from without. Radical jihadist terrorism is metastasizing in every continent. Our magnificent skyscrapers jutted into the sky in proud defiance of this solemn warning—until 9/11. Now it is happening here. Our world has forever changed, and we can no longer afford to think provincially. Perhaps now some may listen more carefully. Perhaps some may throw off the scales from their eyes. Only time will tell.

The scriptures foretell of a time when 10 government leaders will eventually give their power to one who is coming with flowery promises of peace and prosperity. But peace will not come. Instead, tyranny will reign over humankind with more fury than ever before in human history.

The talons of the dragon reach deep, far, and wide into our planet. And his minions have cleverly dressed their doctrines in multifaceted and intriguing forms to pander to greed that leads to violence throughout the planet with cunning deception. The rise in lawlessness and violence and a resulting imposed and cleverly engineered economic upheaval will become the irresistible impetus for the coalescing of these 10 leaders who will give power to the 11th.

We can no longer claim ignorance. We have been duly warned. Our children will pay the price—they will bear the lion's share—for the anguish and misery naturally birthed from our apathy.

"Is it possible or impossible to transmit the experience of those who have suffered to those who have yet to suffer? Can one part of humanity learn from the bitter experience of another or can it not? Is it possible or impossible to warn someone of danger?"[8] Is it possible for America to understand the far-reaching implications of abandoning the Judeo-Christian foundation of our American Revolution?

Judicial Tyranny

Our noble American experiment is showing deep cracks in its foundation as satanic forces continue to weave their tale of peace and prosperity to those who follow this wretched Pied Piper away from our biblical foundation toward a secular, global, socialist-democratic government. It is not just the executive branch of the federal government that is expanding its previsioned parameters set by the Constitution; it is also the judicial branch. The net effect of this dual attack is to bypass the will of the people and rob Congress of its specific role in our government to represent our interests and protect our freedoms.

The absolute standard of truth from which the Bill of Rights was formed is being more than just neglected. It is being opposed and replaced as progressive members of the U.S. Supreme Court have slowly reinterpreted the Constitution through the relativistic Hegelian grid of modern secular existential thought.

This third branch of our government, which was designed explicitly to separate the legislative and executive powers, has overstepped its constitutional boundaries, and jurisprudence has turned into a legislative body by fiat decisions redacted completely outside the constitutional bounds intended by the original framers.

The overwhelming number of justices in every echelon of the judicial system who now perceive their judicial mandates through the tainted spectacles of the Hegelian-materialistic worldview is such that the tripartite system designed by our Founding Fathers is crumbling before our eyes.

The division of power established by them to ensure that one branch does not overpower another and in order for our representative form of government to, in fact, reflect the legal foundation based on the absolute moral principles that were the bedrock of our nation is now being threatened and systematically disassembled. Most Americans are woefully unaware of this great travesty. But even more alarming is that those who have become aware are simply apathetic to these dire developments.

It was the legislative branch of our government that was endowed with the responsibility to redact laws. And for this reason, they are subject to popular vote and must be responsible to the people that elect them. They are to be the voice of we the people. The judicial branch was designed to be free from the political pressure of the popular vote and interpret the laws redacted by Congress through the grid of the U.S. Constitution and the absolute standards enumerated therein. In this way, their duty was to defend the rights of the people by forcing all laws to remain faithful to the immutable Bill of Rights created by our Founding Fathers.

Instead, we now have a new breed of modern judges who have overstepped their duty to interpret law and moved away from the foundation of the Constitution. They are liberally redacting laws by fiat interpretations in order to sidestep the Constitution and enforce their own atheistic and Hegelian agendas on the populace.

They have, in essence, stepped outside their constitutional boundary and brazenly taken over the responsibility of the legislative branch of our government. The clever tactics of the progressivist-socialist agenda is to disguise their crime through clever words that mask their attempt to bypass the will of the people and their voice in Congress. By declaring the Constitution a living document, they freely attempt to sidestep the legal constraints written in the Constitution that ensure the rights of the people and in doing so impose their will on the American people.

This is judicial tyranny of the first order and completely unconstitutional. If we are to survive as a free nation and if our sacred American Revolution is to succeed, then this travesty must be remanded. In this there can be no equivocation.

The great middle class of our nation has long been the anchor that has maintained our freedoms in this country. And their power to oust legislators who would threaten that freedom has, thus far, kept us reasonably free.

Today, this has been circumvented by the power of television and the money that is necessary to run for office. The power of the middle class has also been sidestepped both by unconstitutional executive orders and with the new breed of justices who, in fact, can by judicial fiat, force upon us laws that sidestep our constitutional system ignoring the very basic precepts for which our American Revolution was fought.

Nominating justices to the U.S. Supreme Court who are committed to the strict adherence of the foundational precepts of our Constitution and Declaration of Independence, as designed by our Founding Fathers is of paramount importance if we are to remain free as a people. May God give us the grace to understand the importance of placing strict constructionists to this most important position in our judicial system. If we do not, that may prove to be our greatest Achilles' heel.

Our voices must be heard in Congress. We must do everything in our power to elect presidents who will nominate justices who will defend the letter and intent of the Constitution. We must put pressure on our senators and representatives to approve such appointments. And if they do not, we must vote them out of office. We must push for real election reforms that will bring our votes to real meaning.

As a result of the justices placed on the U.S. Supreme Court during President George W. Bush's terms, the Court for the first time ruled in 2007 in favor of limiting the right to kill the unborn by upholding the law that prohibited partial birth abortions. But it took 35 years since *Roe v. Wade* to finally have the slim majority necessary to uphold the value of human life. Justice Antonin Scalia, the backbone of the justices who defended the Constitution, mysteriously died in 2016, and that margin was gone. The future of our nation was dancing perilously on the edge of a knife. Had Hillary won the presidential election in 2016, there would have been no turning back. But God was gracious to us, and, like a bolt from the blue, Donald Trump won when everyone believed it was going to be an easy victory for Hillary Clinton.

Since the election of Barack Obama as president of the United States in 2008, much has been done to push us down the road to socialism. The future looked bleak, indeed. The progressivist measures he instituted in such overwhelming speed brought us closer to a socialist state than ever before in our American history. The ideology of the forced redistribution of wealth is now being jammed down the throats of Americans with alarming speed and efficiency. Our capitalist, free-market economy was being strangled by specious overregulations that stifled our economy and could lead us into bankruptcy. Our water rights were being robbed from us. Ranchers were unnecessarily killed by federal agents for resisting this usurpation of their rights; it was beginning to sound like the KGB was in America.

Then, out of nowhere, Trump came and surprised the heck out of everyone, including me, and won against all odds in the 2016 presidential election. Had Hillary won, we would have had socialism on steroids for the next eight years. The U.S. Supreme Court would have had a majority of liberal justices with their Hegelian spectacles, and the damage that could have come from that is immeasurable.

In fact, the damage could have been unrecoverable. But God had mercy on us and pulled a miracle out of thin air. President Trump has providentially been able to put not one but two constitutional conservatives on the U.S. Supreme Court, averting the disaster that was looming on the near horizon. Thus far, Trump has begun to undo much that Obama had done with his executive orders and stifling regulations. He is beginning to clean the swamp. But the power of the swamp cannot be underestimated.

The American neo-nobility is not happy with all Trump has undone. He appointed two soundly constitutional conservative justices—Neil Gorsuch and Brett Kavanaugh. For the moment, we have a narrow margin to hold back their globalist plans. Trump passed a tax reform bill that kicked in our economy and will eventually help bring down the national debt.

Trump is wisely strengthening the long-neglected armed forces necessary to keep us protected from the growing menace of our raging planet. Trump has kicked the hornet's nest in the deep Washinton, DC, swamp, and they are going after him with a vengeance. The battle is on. The danger is not yet over. The deep swamp creatures are coming out of the murky waters with knives in their mouths.

Our children are still being indoctrinated. Our religious freedoms are still being eroded as our voices have been removed by violence from the institutions of higher learning. Our mass media have become so brazenly partisan that for all practical purposes, they are simply a propaganda wing of the globalists and socialists. Hatred drips from their raging mouths like I have never seen before in my 66 years of living as they try to destroy Trump's agenda and besmirch his integrity. In the past, their tactic was subtle and deceptive. Today, they have thrown all caution to the wind, and their elitist arrogance is surpassed only by their ignorance of the true nature of the beast they peddle.

There is no substantial ideological difference between our American mainstream media and Stalin's government-controlled media during the Soviet Union. It is a monolithic echo chamber that spews out leftist ideology and rhetoric without dissimulation. The once stately and fair-minded reporters who gave us unbiased news are a thing of the past. They have become extinct.

We need only to glance at Europe to see the end result of such socialist policies. Greece is on the edge of becoming economically insolvent, and Italy, Spain, and Portugal are right behind it. If we continue down this road, so will we.

Socialism and Globalization – The Threat from the West

There seems to be a pattern emerging in America. The stage is being set for the blending of our nation with other nations of the world in preparation for a one-world government. Most of Europe has already been swallowed up by the socialist agenda of the promoters of globalization. The oligarchies of the West will attempt to unite all nations under their economic grip.

The foundational precepts of Samuel Rutherford's book *Lex Rex* that formed our country will be replaced by a system of laws based on a socialistic and humanistic foundation. Our noble American experiment birthed by the Judeo-Christian worldview will one day eventually succumb to Apollyon.

But ours is the duty, privilege, and responsibility to resist the tyrannical forces afoot. Ours is the most difficult task of any generation, for more is at stake now than ever before as we reach the culmination of our epoch.

Apathy will be the lethal blow that brings an end to the freedoms we now enjoy. And we will have no one to blame but ourselves if we do not learn to lovingly and intelligently defend our beloved nation against the coming night.

In our past, many turned a blind eye to colonialism and imperialism out of selfish greed. But be not deceived; we will reap what we sow. Our heartless abuse of the less fortunate will be judged. Of that you can be sure. Solzhenitsyn understood that very well:

> However hidden it may be from human gaze, however un-expected for the practical mind, there is sometimes a direct link between the evil we cause to others and the evil which suddenly confronts us. Pragmatists may explain this link as a chain of natural causes and effect. But, those of us who are more inclined to a religious view of life will immediately perceive a link between sin and punishment. It can be seen in the history of every country. Today's generation has had to pay for the shortcomings of their fathers and grand-fathers, who blocked their ears to the lamentations of the world and closed their eyes to its miseries and disasters.[9]

Hatred and greed are curved blades that eventually cut the arms that wield them. Will we turn a blind eye to the same forces now functioning with impunity on our planet? Where will our line rest within our bosoms? Will we block our ears to the lamentations of our oppressed brothers throughout our world, both born and unborn? Will we fall off the side of the cliff called collectivism or supernationalism? Either expression of greed leads to totalitarianism and the loss of freedom.

Only the Judeo-Christian worldview encompasses the proper balance of community and individual dignity. We are each infinitely valuable, but we are not an island. We are to selflessly bear one another's burdens. That is true and balanced community.

> *Bear ye one another's burden and so fulfil the law of Christ.*
> —Gal. 6:2 KJV

Here is where the rubber meets the road. True community cannot take place without the assistance of the Spirit of God within us in order to help us move that line in our bosoms toward the side of good. Our selfish human nature must be set aside, and we must seek to be selfless, preferring to meet our neighbors' needs before our own lusts.

Let love be without hypocrisy. Abhor what is evil; cling to what is good. Be devoted to one another in brotherly love; give preference to one another in honor.

—Rom. 12:9–10

The Judeo-Christian worldview is the only worldview that is balanced in its approach to the true nature of humans and society. We are not idealists, nor are we pessimists. We are true realists.

Those who champion collectivism are idealists who erroneously believe that the good in humanity is sufficient to establish a classless utopian society with absolute equality. But the evil in humankind is such that tyrants and despots eventually subjugate the masses, betraying the naïve communist ideal of absolute equality for all people.

The reality is that the line for each person is different, and thus there is no such thing as absolute equality. **Some people are simply human monsters.** To deny this truth is to grant those monsters the opportunity to wield their wicked schemes. To deny this truth is to deny reality. To deny this truth is to choose tyranny. To deny this truth is to neglect our rightful place in society as the preservers of justice and peace. To deny this truth is to grant evil the road to victory. We must not blind ourselves to this fundamental truth, or we will condemn our posterity, and they will one day hold us accountable.

The proponents of supernationalism are hyperindividualists who are deceived into thinking they are superior in one way or another to the rest of humanity. These are the pessimists who claim that most people are inferior to a select few and therefore should be subjugated to serve the elite. Colonialism and imperialism are the tools of those who peddle supernationalism. It is the greedy, ever-growing obsession with annexation of new victims to exploit. Its hunger can never be abated. Its thirst for riches and power can never be quenched. It is the most horrendous fruit of the fall of humankind.

Avarice is ever the beating heart of the promoters of supernationalism and colonialism. These are the monsters filled with arrogance and self-importance whose narcissistic insolence and pride have ever mimicked the foremost enemy of God; that is, Satan, the opposer and impostor. But their end will be the Lake of Fire. For One is coming from the root of Jesse, from the seed of David, and "the government shall be upon his shoulder" (Isa. 9:6 KJV).

The time will come when our second earth will have run its course and reached the fullness of its indignation. The King of the North will bring war and death upon our planet. For seven years, our second earth will suffer the death pangs of the seven-year tribulation period—the time of Jacob's trouble, the time of Achor. He will make war with the saints and cause evil to prosper. Daniel saw that day:

> *Then the king will do as he pleases, and he will exalt and magnify himself above every god and will speak monstrous things against the God of gods; and he will prosper until the indignation is finished, for that which is decreed will be done. He will show no regard for the gods of his fathers or for the desire of women, nor will he show regard for any other god; for he will magnify himself above them all. But instead he will honor a god of fortresses, a god whom his fathers did not know; he will honor him with gold, silver, costly stones and treasures. He will take action against the strongest of fortresses with the help of a foreign god; he will give great honor to those who acknowledge him and will cause them to rule over the many, and will parcel out land for a price.*
>
> —Dan. 11:36–39

This King of the North is he who comes on the wings of abomination to make all lands desolate. But in the middle of that week of years, he will stand outside Jerusalem to conquer it with the help of the demon who calls himself the god of fortresses or the god of war. Apollyon shall rise from the great abyss and give aid to the King of the North (Rev. 9:11).

Azazel, who taught humanity the art of war, will rise from his punishment along with 200 other demons who led the second insurrection in the first earth. He is the god of fortresses who will aid in the fourth insurrection.

> *And he will make a firm covenant with the many for one week,*
> *but in the middle of the week he will put a stop to sacrifice and*
> *grain offering; and on the wing of abominations will come one*
> *who makes desolate, even until a complete destruction, one*
> *that is decreed, is poured out on the one who makes desolate.*
>
> —Dan. 9:27

In that day, in the middle of the week of years, he will kill the two lampstands, the two olive trees that God will send in order to prepare His chosen to find the great succoth in the wilderness. They will minister to the chosen of God for the first three and a half years of the Great Tribulation. But in the middle of the week, the King of the North will conquer Jerusalem and kill the two prophets.

For three and a half days, Elijah and Enoch will lay dead, killed by Apollyon (Azazel) in Jerusalem. There the King of the North will claim godhood inside the Holy of Holies in God's third temple. This is the fourth insurrection and marks the beginning of God's wrath through the judgment of the seventh trumpet. From that day forward, the seven bowls of God's wrath will be poured on the kingdom of the usurper to atone for the earth.

> *"And I will grant authority to my two witnesses, and they*
> *will prophesy for twelve hundred and sixty days, clothed*
> *in sackcloth." These are the two olive trees and the two*

lampstands that stand before the Lord of the earth. And if anyone wants to harm them, fire flows out of their mouth and devours their enemies; so if anyone wants to harm them, he must be killed in this way. These have the power to shut up the sky, so that rain will not fall during the days of their prophesying; and they have power over the waters to turn them into blood, and to strike the earth with every plague, as often as they desire.

When they have finished their testimony, the beast that comes up out of the abyss will make war with them and overcome them and kill them. And their dead bodies will lie in the street of the great city which mystically is called Sodom and Egypt, where also their Lord was crucified. Those from the peoples and tribes and tongues and nations will look at their dead bodies for three and a half days and will not permit their dead bodies to be laid in a tomb. And those who dwell on the earth will rejoice over them and celebrate; and they will send gifts to one another, because these two prophets tormented those who dwell on the earth.

But after the three and a half days, the breath of life from God came into them, and they stood on their feet; and great fear fell upon those who were watching them. And they heard a loud voice from heaven saying to them, "Come up here." Then they went up into heaven in the cloud, and their enemies watched them. And in that hour, there was a great earthquake, and a tenth of the city fell; seven thousand people were killed in the earthquake, and the rest were terrified and gave glory to the God of heaven.

—Rev. 11:3–13

Then before the eyes of the watching world, the two lampstands will resurrect and rise up to heaven before their eyes. The occult doctrine, "as above so below," is here laid bare and defeated. At the same time the King of the North attacks Jerusalem, Lucifer is attacking heaven. And there was a war in heaven, but Lucifer was not strong enough. And Michael the Archangel threw him to the earth.

And there was war in heaven, Michael and his angels waging war with the dragon. The dragon and his angels waged war, and they were not strong enough, and there was no longer a place found for them in heaven. And the great dragon was thrown down, the serpent of old who is called the devil and Satan, who deceives the whole world; he was thrown down to the earth, and his angels were thrown down with him. Then I heard a loud voice in heaven, saying,

"Now the salvation, and the power, and the kingdom of our God and the authority of His Christ have come, for the accuser of our brethren has been thrown down, he who accuses them before our God day and night. And they overcame him because of the blood of the Lamb and because of the word of their testimony, and they did not love their life even when faced with death. For this reason, rejoice, O heavens and you who dwell in them. Woe to the earth and the sea, because the devil has come down to you, having great wrath, knowing that he has only a short time."

And when the dragon saw that he was thrown down to the earth, he persecuted the woman who gave birth to the male child. But the two wings of the great eagle were given to the woman, so that she could fly into the wilderness to her place, where she was nourished for a time and times and half a time, from the presence of the serpent. And the serpent poured water like a river out of his mouth after the woman, so that he might cause her to be swept away with the flood. But the earth helped the woman, and the earth opened its mouth and drank up the river which the dragon poured out of his mouth. So the dragon was enraged with the woman, and went off to make war with the rest of her children, who keep the commandments of God and hold to the testimony of Jesus.

—Rev. 12:7–17

Hear my words, O Judah. When Elijah comes, listen to his words—your lives depend upon it. Flee to the wilderness where the Eagle spreads His wings. There and only there shall you find the Bread of Life, the Living Water, the Rock of Ages, and the Light of God to protect you for three and a half years until the Indignation is complete. It is the Great Succoth, the ark of the second earth.

And so shall His Chosen resurrect from the Great Succoth in the wilderness after three and a half years from that day. When the One, the Anointed One, who is decreed, the Lion of Judah comes to destroy the armies of the Gentile nations gathered at Megiddo to finally end the Indignation. God shall allure you to the wilderness, the hiding place, where you shall see Him face to face.

"Therefore, behold, I will allure her,
Bring her into the wilderness
And speak kindly to her.
"Then I will give her her vineyards from there,
And the valley of Achor as a door of hope.
And she will sing there as in the days of her youth,
As in the day when she came up from the land of Egypt.
"It will come about in that day," declares the Lord,
"That you will call Me Ishi
And will no longer call Me Baali.
"For I will remove the names of the Baals from her mouth,
So that they will be mentioned by their names no more.
"In that day I will also make a covenant for them
With the beasts of the field,
The birds of the sky

And the creeping things of the ground.
And I will abolish the bow, the sword and war from the land,
And will make them lie down in safety.
"I will betroth you to Me forever;
Yes, I will betroth you to Me in righteousness and in justice,
In lovingkindness and in compassion,
And I will betroth you to Me in faithfulness.
Then you will know the LORD.

"It will come about in that day that I will respond," declares
the LORD.
"I will respond to the heavens, and they will respond to the earth,
And the earth will respond to the grain, to the new wine and
to the oil,
And they will respond to Jezreel.
"I will sow her for Myself in the land.
I will also have compassion on her who had not obtained
compassion,
And I will say to those who were not My people,
'You are My people!'
And they will say, 'You are my God!'"

—Hosea 2:14–23

In that day, Azazel and his demons, along with the King of the North and his Sorcerer, shall be the firstfruits thrown into the lake of fire. And Lucifer shall be cast in chains into the abyss for a thousand years.

And the beast was seized, and with him the false prophet who
performed the signs in his presence, by which he deceived
those who had received the mark of the beast and those who
worshiped his image; these two were thrown alive into the
lake of fire which burns with brimstone.

—Rev. 19:20

In that day, the root of Jesse shall have the government upon His shoulders. He shall rule with the iron scepter of Judah from Jerusalem. He will not need a cabinet of advisors. He will not need surveillance equipment and sophisticated technology to spy upon us. He will know the truth, the whole truth, the hidden truth, and He will judge with equity. In that glorious day, justice shall prevail throughout our planet. The government shall rest solely upon His shoulders.

The weak shall not be trodden, and the poor shall not be oppressed. The wicked merchants of the world who fed off the misfortune of those they robbed through usury and many other hostile schemes to defraud the powerless shall be judged. Gone shall be the judges who take bribes and the politicians who sell our rights to vested interests.

Righteousness shall rule the planet, and peace shall finally come to humankind. He shall be the Son of David, and with the rod of iron as His scepter of Judah and the roar of the mighty Lion of Judah, He shall strike the earth. He will slay the wicked, and righteousness shall finally flourish.

The matrix of reality shall shift from the present paradigm of the survival of the fittest through predation to that of peace, justice, and righteousness. The wolf shall lie down with the lamb, and nature will no longer be cruel.

Then a shoot will spring from the stem of Jesse,
And a branch from his roots will bear fruit.
The Spirit of the LORD will rest on Him,
The spirit of wisdom and understanding,
The spirit of counsel and strength,
The spirit of knowledge and the fear of the LORD.
And He will delight in the fear of the LORD,
And He will not judge by what His eyes see,
Nor make a decision by what His ears hear;
But with righteousness He will judge the poor,

And decide with fairness for the afflicted of the earth;
And He will strike the earth with the rod of His mouth,
And with the breath of His lips He will slay the wicked.
Also righteousness will be the belt about His loins,
And faithfulness the belt about His waist.

And the wolf will dwell with the lamb,
And the leopard will lie down with the young goat,
And the calf and the young lion and the fatling together;
And a little boy will lead them.
Also the cow and the bear will graze,
Their young will lie down together,
And the lion will eat straw like the ox.
The nursing child will play by the hole of the cobra,
And the weaned child will put his hand on the viper's den.
They will not hurt or destroy in all My holy mountain,
For the earth will be full with the knowledge of the Lord
As the waters cover the sea.

Then in that day
The nations will resort to the root of Jesse,
Who will stand as a signal for the peoples;
And His resting place will be glorious.

—Isa. 11:1–10

Our third earth will be transformed back to the way God intended it to be during the first earth. Israel will come out of the ark of the second earth underneath the Eagle's wings, and the Great Succoth will be remembered from that day forward as God's miraculous protection of those who will become His priests in the kingdom of David.

O God, You have rejected us. You have broken us;
You have been angry; O, restore us.
You have made the land quake, You have split it open;

Heal its breaches, for it totters.
You have made Your people experience hardship;
You have given us wine to drink that makes us stagger.
You have given a banner to those who fear You,
That it may be displayed because of the truth. Selah.
That Your beloved may be delivered,
Save with Your right hand, and answer us!

God has spoken in His holiness:
"I will exult, I will portion out Shechem and measure out
the valley of Succoth."

—Ps. 60:1–6

In that day, the valleys will quake, the mountains will crumble, the valleys will be lifted up, islands will fall into the sea. And the seven continents will become one again. Jerusalem will be the center of the earth. It will be known as the City of Truth where God will dwell among us and rule with righteousness through the Iron Scepter of Judah. Justice will prevail for all humankind. The meek will inherit the earth.

Behold, My servant will prosper,
He will be high and lifted up and greatly exalted.
Just as many were astonished at you, My people,
So His appearance was marred more than any man
And His form more than the sons of men.
Thus He will sprinkle many nations,
Kings will shut their mouths on account of Him;
For what had not been told them they will see,
And what they had not heard they will understand.

—Isa. 52:13–15

As we see the end days approaching, we can perceive more clearly the globalist plans of the occult through collectivism taking shape on one side of the battle, while on the other side, the globalist plans of Islam continue to progress through the same supernationalist ideology. Both of these corrupted forms of true and balanced nationalism will lead to global tyranny. Which one will gain the upper hand, only time will tell. But in either case, the Enemy of Man will rule the planet.

So what, then, should be our duty today? If we know that in the end the Antichrist will rule, what are we to do? We have the same duty that has faced all believers in every age throughout history—to stand for righteousness and truth despite the overwhelming odds of victory. To do anything less is to disobey God and be a coward. Our duty is to proclaim the truth, promote justice and righteousness, and battle wickedness until our dying breath, and then leave the consequences to God.

The truth is that we are the exception in human history. No other generation in the world, except perhaps at the inception of Israel, has ever existed with the freedoms that we enjoy. The problem is that we have taken them for granted and are so engrossed in our pursuit of pleasure that we cannot see them crumbling beneath our feet. This boat is sinking, and the storm is almost upon us. If we hope to endure the storm, we must repair our ship as quickly as possible.

Will history remember our generation as the one that allowed these cherished freedoms so longed for by all other generations to be taken from us? That depends on what you are willing to do today.

Notes

1. Henry M. Christman, ed., *Essential Works of Lenin: "What Is to Be Done" and Other Works* (New York: Dover Publications, 1987), 236–237.
2. Aleksandr I. Solzhenitsyn, *The Gulag Archipelago* (New York: Harper & Row, 1973), 68.
3. Francis Schaeffer, *A Christian Manifesto* (Westchester, IL: Crossway Books, 1982), 60.

4. Aleksandr Solzhenitsyn, *Stories and Prose Poems* (New York: Farrar, Straus and Giroux, 1971), 131.
5. Aleksandr Solzhenitsyn, *Warning to the West* (New York: Ferrar, Straus and Giroux, 1976), 33–34 .
6. "Inaugural Address of John Adams," March 4, 1979, Philadelphia, PA, *Yale Law School, The Avalon Project*, http://avalon.law.yale.edu/18th_century/adams.asp.
7. Aleksandr Solzhenitsyn, *Warning to the West*, 52–53.
8. Ibid.
9. Ibid., 140.

ABOUT THE AUTHOR

Henry Patiño was born in Buenos Aires, Argentina, and raised in Cuba. After Fidel Castro imprisoned his father for political reasons, Patiño's family immigrated to the United States in 1961. He came to the Christian faith on Easter 1970. He holds a bachelor of science degree in theology and is an ordained evangelical minister. Rev. Patiño has dedicated his life to the spread of the good news of the gospel of Christ, the mentoring and discipling of others in the body of Christ, and the defense of the Word of God.

He has founded and led several youth ministries, two churches, a home for unwed mothers, and four crisis pregnancy centers throughout South Florida.

He has served as associate pastor, pastoral counselor, and senior pastor in several evangelical Christian churches.

Rev. Patiño has also tirelessly devoted himself to the sanctity of human life as one of the early pioneers of the movement. He has been successful in helping to unite various Christian denominations throughout South Florida in this common and vital cause in the evangelical community.

He was one of the founding board members of the Miami chapter of the Christian Action Council, serving first as vice president and later as president. The Christian Action Council was a national pro-life organization founded by the late Dr. Francis Schaeffer and former Surgeon General Dr. C. Everett Koop with the purpose of (1) educating the evangelical community in the issues of abortion, infanticide, and euthanasia, (2) establishing crisis pregnancy centers funded and supported by evangelical Christian churches, and (3) developing positive, nonviolent political action in local communities, aimed at opposing legislation that threatened our God-given right to life.

Patiño was also instrumental in the 1982 March for Life Pro-Life Conference in South Florida that brought such notables as Dr. Francis Schaeffer, Franky Schaeffer IV, Phyllis Schaffley, Joe Scheidler, Phil Kaeggy, Dr. D. James Kennedy, Pat Robertson, Cal Thomas, and John Whitehead. The march-conference was the first national pro-life outreach to the evangelical community in the United States organized by Coral Ridge Presbyterian Church of Ft. Lauderdale through the instigation of Jean Emond.

Rev. Patiño also founded and served as chairman of the board of the South Florida Coalition for Life, an umbrella group that united all the pro-life groups in South Florida and was comprised of both Roman Catholics and evangelicals.

He has been guest speaker at many crisis pregnancy center fundraisers and frequented television and radio programs on which he successfully debated pro-abortion advocates.

Throughout most of his years of service to the Lord Jesus Christ and to the body of Christ, Patiño has managed to fulfill his duties as a fireman-paramedic for the City of Miami Beach, choosing to "build tents" for a living. From 1987 to 1991, he served as chaplain for the Miami Beach Fire Department.

Aside from his duties as a fireman-paramedic, Patiño devotes much of his time to writing, mentoring, and overseas medical missions. The scope of his books includes a wide range of disciplines that he has carefully woven into his desire to create intelligent apologetics for the Judeo-Christian worldview.